Occupational Health and Safety
in the Care and Use of Nonhuman Primates

Committee on Occupational Health and Safety
in the Care and Use of Nonhuman Primates

Institute for Laboratory Animal Research
Division on Earth and Life Studies

NATIONAL RESEARCH COUNCIL
OF THE NATIONAL ACADEMIES

THE NATIONAL ACADEMIES PRESS
Washington, D.C.
www.nap.edu

THE NATIONAL ACADEMIES PRESS 500 Fifth Street, N.W. Washington, DC 20001

NOTICE: The project that is the subject of this report was approved by the Governing Board of the National Research Council, whose members are drawn from the councils of the National Academy of Sciences, the National Academy of Engineering, and the Institute of Medicine. The members of the committee responsible for the report were chosen for their special competences and with regard for appropriate balance.

This study was supported by contract number N01-OD-4-2139, Task Order No. 73 and purchase order number 263-MD-005750 from the National Institutes of Health (NIH), Department of Health and Human Services (DHHS); grant number R13RR15165 from the National Center for Research Resources (NCRR), NIH, DHHS; purchase order number 0000065415 from the National Center for Infectious Diseases, Centers for Disease Control and Prevention (CDC), DHHS; purchase order number 0000166565 from the National Institute of Occupational Safety and Health, CDC, DHHS; contract number 223-93-1025 from the Food and Drug Administration, DHHS. Financial support was also provided by the Association of Primate Veterinarians, Merck Research Laboratories, and the Elizabeth R. Griffin Foundation.

Core support for the Institute for Laboratory Animal Research is provided by the Division of Comparative Medicine, NCRR, NIH through grant number P40RR11611; the National Science Foundation through grant number DBI-9805555; the US Army Medical Research and Development Command, which serves as the lead agency for combined US Department of Defense funding also received from the Human Systems Division of the US Air Force Systems Command, Armed Forces Radioiology Research Institute, Uniformed Services University of the Health Sciences, and US Naval Medical Research and Development Command through grant number DAMD-98-1-8275; and Merck Research Laboratories.

International Standard Book Number 0-309-08914-X (Book)
International Standard Book Number 0-309-50779-0 (PDF)
Library of Congress Catalog Card Number 200310660

Additional copies of this report are available from the National Academies Press, 500 Fifth Street, N.W., Lockbox 285, Washington, DC 20055; (800) 624-6242 or (202) 334-3313 (in the Washington metropolitan area); Internet, http://www.nap.edu

Printed in the United States of America

THE NATIONAL ACADEMIES

Advisers to the Nation on Science, Engineering, and Medicine

The **National Academy of Sciences** is a private, nonprofit, self-perpetuating society of distinguished scholars engaged in scientific and engineering research, dedicated to the furtherance of science and technology and to their use for the general welfare. Upon the authority of the charter granted to it by the Congress in 1863, the Academy has a mandate that requires it to advise the federal government on scientific and technical matters. Dr. Bruce M. Alberts is president of the National Academy of Sciences.

The **National Academy of Engineering** was established in 1964, under the charter of the National Academy of Sciences, as a parallel organization of outstanding engineers. It is autonomous in its administration and in the selection of its members, sharing with the National Academy of Sciences the responsibility for advising the federal government. The National Academy of Engineering also sponsors engineering programs aimed at meeting national needs, encourages education and research, and recognizes the superior achievements of engineers. Dr. Wm. A. Wulf is president of the National Academy of Engineering.

The **Institute of Medicine** was established in 1970 by the National Academy of Sciences to secure the services of eminent members of appropriate professions in the examination of policy matters pertaining to the health of the public. The Institute acts under the responsibility given to the National Academy of Sciences by its congressional charter to be an adviser to the federal government and, upon its own initiative, to identify issues of medical care, research, and education. Dr. Harvey V. Fineberg is president of the Institute of Medicine.

The **National Research Council** was organized by the National Academy of Sciences in 1916 to associate the broad community of science and technology with the Academy's purposes of furthering knowledge and advising the federal government. Functioning in accordance with general policies determined by the Academy, the Council has become the principal operating agency of both the National Academy of Sciences and the National Academy of Engineering in providing services to the government, the public, and the scientific and engineering communities. The Council is administered jointly by both Academies and the Institute of Medicine. Dr. Bruce M. Alberts and Dr. Wm. A. Wulf are chair and vice chair, respectively, of the National Research Council.

www.national-academies.org

COMMITTEE ON OCCUPATIONAL HEALTH AND SAFETY IN THE CARE AND USE OF NONHUMAN PRIMATES

Frederick A. Murphy *(Co-Chairman)*, University of California - Davis, School of Veterinary Medicine, Davis, California
Jeffrey A. Roberts *(Co-Chairman)*, University of California - Davis, California National Primate Research Center, Davis, California
Kathryn A.L. Bayne, Association for Assessment and Accreditation of Laboratory Animal Care International, Rockville, Maryland
James L. Blanchard, Tulane National Primate Research Center, Covington, Louisiana
Thomas J. Ferguson, University of California - Davis, Davis, California
Lisa J. Flynn, US Public Health Service Commissioned Corps, US Food and Drug Administration, Rockville, Maryland
Jack Geissert, Wyeth BioPharma Division, Andover, Massachusetts
Julia K. Hilliard, Georgia State University, Department of Biology, Atlanta, Georgia
Michael Kiley, US Department of Agriculture, Beltsville, Maryland
Clarence J. Peters, University of Texas Medical Branch, Department of Microbiology, Immunology, and Pathology, Galveston, Texas
Benjamin J. Weigler, Fred Hutchinson Cancer Research Center, Seattle, Washington

Consultant

David S. Davenport, Michigan State University, Kalamazoo, Michigan

Staff

Joanne Zurlo, Director
Marsha Barrett, Project Assistant
Kathleen Beil, Administrative Assistant
Ralph Dell, Associate Director
Norman Grossblatt, Editor
Jennifer Obernier, Program Officer
Susan Vaupel, Editor

INSTITUTE FOR LABORATORY ANIMAL RESEARCH COUNCIL

Preface

The publication in 1997 of the Institute for Laboratory Animal Research (ILAR) report *Occupational Health and Safety in the Care and Use of Research Animals* provided an excellent reference for the development of occupational health and safety programs in the animal research setting. The diversity of species and potential hazards encountered in animal care and use programs required a broad view of many topics and by necessity, limited the depth of any particular subjects in the report. The care and use of nonhuman primates in the research setting presents a number of challenges to facility management. These challenges include specific hazards unique to some primate species and the need for guidance in risk assessment and management. This report was generated in response to that need and to specific events that took place in the same year as the first ILAR report.

On October 29, 1997, a research assistant at Yerkes Regional Primate Research Center was splashed in the eye with an unidentified body fluid from a nonhuman primate and later died from encephalitis caused by B virus (formerly called Cercopithecine herpesvirus 1). This incident confirmed the suspicion that B virus infection can be acquired through mucosal contact, in addition to the more common exposures through bites, scratches, and needle sticks. Following the incident at Yerkes, the National Institute for Occupational Safety and Health (NIOSH) conducted a limited review of policies and procedures related to working with nonhuman primates at various National Primate Research Centers. NIOSH recommended that "goggles and face shields be worn when working with nonhuman primates." Many investigators and caretakers expressed concern that these protective devices would obstruct vision and would be difficult to wear, leading to an increase in bites, scratches, and needles sticks due to decreased vision. There was also the concern that workers would not adhere to the NIOSH recommendation. Therefore, one specific intent of this report has been to address these recommendations and concerns.

Infectious agents represent only one of the hazards present in nonhuman-primate animal care and use programs. The size, strength, and intelligence of many primate species can result in unique hazards associated with animal care and management. In addition, these same taxonomic attributes may require heavy caging and support equipment that can present ergonomic hazards to the employee.

The Committee was asked to identify the hazards associated with using nonhuman primates in research, assess the degree of risk of these hazards, and suggest options for managing the risks including engineering controls, personal protective equipment, facilities design, and worker training. The committee was also asked to make recommendations for institutional management of workers after suspected exposure to infectious agents.

The Committee approached this task by focusing on major hazards and risks to workers at all nonhuman-primate research facilities. Recognizing that the level of risk associated with a hazard is dependent on numerous factors that vary from institution to institution, the Committee on Occupational Health and Safety in the Care and Use of Nonhuman Primates put forth in this report a programmatic structure for assessing and managing risk at different kinds of institutions. In addition to providing this structure and discussing the elements necessary for the successful implementation of an occupational health and safety program, the Committee reviewed specific recommendations on the use of personal protective equipment and the medical management of exposed workers.

This report has been reviewed in draft form by individuals chosen for their diverse perspectives and technical expertise, in accordance with procedures approved by the NRC's Report Review Committee. The purpose of this independent review is to provide candid and critical comments that will assist the institution in making its published report as sound as possible and to ensure that the report meets institutional standards for objectivity, evidence, and responsiveness to the study charge. The review comments and draft manuscript remain confidential to protect the integrity of the deliberative process. We wish to thank the following individuals for their review of this report:

Patrick Breysse, Johns Hopkins Bloomberg School of Public Health, Baltimore, Maryland

Thomas Butler, Southwest Foundation for Biomedical Research, San Antonio, Texas

Roy Henrickson, Private Consultant, Point Richmond, California

Nicholas Lerche, California National Primate Research Center, Davis, California

Keith Mansfield, New England National Primate Research Center, Southborough, Massachusetts

James Schmitt, National Institutes of Health, Bethesda, Maryland

Ara Tahmassian, University of California San Francisco, San Francisco, California

Stuart Zola, Yerkes National Primate Research Center, Atlanta, Georgia

Although the reviewers listed above have provided many constructive comments and suggestions, they were not asked to endorse the conclusions or recommendations, nor did they see the final draft of the report before its release. The review of this report was overseen by W. Emmett Barkley, Howard Hughes Medical Institute, Chevy Chase, Maryland. Appointed by the National Research Council, he was responsible for making certain that an independent examination of this report was carried out in accordance with institutional procedures and that all review comments were carefully considered. Responsibility for the final content of this report rests entirely with the authoring committee and the institution.

Since this report will undoubtedly be updated in the future, the members of the Committee ask that comments, corrections, and ideas for future studies be sent to the Institute for Laboratory Animal Research, The National Academies, 500 Fifth Street, NW, Washington, DC 20001.

Frederick A. Murphy, *Co-Chairman*
Jeffrey A. Roberts, *Co-Chairman*
Committee on Occupational Health and Safety in the Care and Use of Nonhuman Primates

Contents

Occupational Health and Safety
in the Care and Use of
Nonhuman Primates

Executive Summary

The Committee on Occupational Health and Safety in the Care and Use of Nonhuman Primates was appointed by the National Research Council (NRC) in response to requests from the National Institutes of Health, the Centers for Disease Control and Prevention, and the Food and Drug Administration to address the risks associated with occupational exposure to nonhuman primates and suggest practical and efficacious ways of minimizing these risks. Specifically, the committee was asked to:

1. Identify hazards associated with using nonhuman primates in research.
2. Assess the degree of risk of these hazards.
3. Suggest options for managing the risks including engineering controls, personal protective equipment, and worker training.
4. Outline the institutional management of workers after a suspected occupational exposure.
5. Provide sample illustrative occupational health and safety plans for personnel working in large and small nonhuman-primate facilities.

The Committee on Occupational Health and Safety in the Care and Use of Nonhuman Primates drew from the experiences of a number of experts, including infectious disease clinicians, primate veterinarians, primate caregivers, and occupational health professionals during a workshop held at the outset of the project (See Appendix A for a list of participants). These experts agreed that the most effective way to identify and manage hazards associated with nonhuman primates is through the development and implementation of an institutionally specific occupational health and safety program (OHSP). This report discusses in detail the building blocks of a successful OHSP, namely identification of hazards, risk assessment, identification of applicable safety regulations, risk management, and personnel training. It also emphasizes the importance of a strong institutional commitment to an OHSP and the clear delegation of responsibility, authority, and accountability at all stages of development, implementation, evaluation, and re-evaluation of the OHSP.

The National Research Council developed a document on occupational health and safety for animal research facilities (NRC 1997), which has served as a guide for the management of an OHSP and has provided a foundation for the development of an institutional OSHP where none exists. The present report attempts to aid in the development or improvement of OHSP at nonhuman-primate facilities or facilties that use nonhuman-primate blood or tissue and is not intended to duplicate the scope or content of the previous document. Rather, its goal is to complement that publication and expand on topics that are particularly relevant or specific to facilities where nonhuman-primate species are housed. This report has also attempted to address the meaning and implications of uncertainty in risk management.

This report is intended as a reference for vivarium managers, veterinarians, researchers, safety professionals, and any other persons who are involved in developing or implementing an OHSP dealing with nonhuman primates. The diversity of institutions, research programs, and animal colonies makes it impossible to encompass all the details of a complete institutional OHSP in this report. Instead, it attempts to list the important features of an OHSP and provide the tools necessary for informed decision-making in developing an optimal program that meets all particular institutional needs.

The Committee identified and assessed numerous risks, infectious and noninfectious, of working with nonhuman primates or their blood or tissues. Significant risks included ergonomic injuries and illnesses caused by shigella, tuberculosis, and B virus infections. These risks can be effectively dealt with using a layered approach to exposure control. Engineering controls are an essential mode of exposure/injury control and include facility design and specialized equipment such as biosafety cabinets. Work practices within the facility provide another modality in exposure/injury control, but can be most important. Development of standard operating procedures that are universally followed and are integrated into employee training can effectively mitigate many hazards.

Another important element in exposure/injury control is the use of personal protective equipment (PPE). PPE for use in nonhuman-primate facilities should minimally include dedicated clothing, gloves, and mask. The Committee stresses that PPE should only serve as a safety net if engineering and work practices should fail. But in light of the potentially fatal risks associated with B virus and other viral exposures, the appropriate use of PPE is a particularly important issue. The Committee concluded that because of the risk of B virus infection, the use of eye and face protection should be mandatory for individuals working with macaques. The Committee also recommended that eye and face protection be used when working with any Old World primate, due to the potential for infection by other primate viruses such as simian immunodeficiency virus. For other nonhuman-primate species, the Committee recommends that the use of eye and face protection be determined locally, based on risk assessment and management processes outlined in this report.

Appropriate medical care after a suspected occupational exposure to a zoonotic pathogen is another area where specific guidance has been lacking in spite of various federal regulations and guidelines. The Committee determined that the first and often most critical step in developing an OHSP is the establishment of a relationship with a pre-designated occupational health care provider. Involving the designated medical providers in determining exposure risks before an incident occurs may lead to quicker and more efficacious post-exposure management. In this report, the Committee makes specific recommendations for medical management following exposure to or injury from nonhuman primates.

The field of occupational health and safety constantly changes, especially as it pertains to biomedical research. The emergence of new hazards presents diverse challenges to employers who must ensure the safety of their employees. New infectious hazards are of particular importance at nonhuman-primate facilities. For example, the discovery that B virus can be transmitted via a splash on a mucous membrane raises new concerns that must be addressed, as does the discovery of the Reston strain of Ebola virus in import quarantine facilities in the United States. The risk of such infectious hazards is best managed through a flexible and comprehensive OHSP that can identify and mitigate potential hazards. It is incumbent on those responsible for nonhuman-primate research facilities, from the senior institutional officer to the facility manager to line supervisors, to develop, improve, and implement such a program.

1

Introduction and Overview

The field of occupational health and safety (OHS) has become a topic of increasing importance over the last 30 years. The establishment of the Occupational Safety and Health Administration (OSHA) in 1970 reflected the recognition that safety in the workplace is a basic expectation for all employees. Originally addressing concerns in industry and hazards associated with mechanical injury, the field of occupational health and safety has expanded to almost every workplace environment, from the office to the airplane, as well as to the laboratory and the vivarium.

The issue of OHS is clearly relevant to biomedical research and extends to the use of animals in biomedical research (NRC 1997). As with any laboratory environment, facilities that house nonhuman primates have a variety of mechanical, chemical and infectious hazards. With new developments in research technology, there is the potential for a variety of real and perceived unique hazards that could make the management of OHS in this type of workplace a challenging endeavor.

A review of safety records pertinent to animal care occupations demonstrates that many of the health hazards encountered when working with nonhuman primates are not unique. A survey of accidental injuries associated with nonhuman primates at two national primate research centers documented a list of occupational injuries including bites, animal-inflicted scratches, needle sticks, cuts, and mucous membrane exposures (bin Zakaria and others 1996). Similarly, surveys of injury reports of employees at veterinary clinics indicate that the most common injuries were animal bites and kicks, needle sticks, and crushing injuries, and that over 50% of the respondents had at least one injury incident over a 3-year period (Poole and others 1998, 1999). The important message from these comparisons is that any animal care occupation has a wide variety of workplace hazards. Once the hazards are identified, the same safety-driven approaches that are used to reduce employee risk in other fields of animal care and use, as well as in other workplace settings, are likewise applicable to people working with nonhuman primates.

OCCUPATIONAL HEALTH AND SAFETY PROGRAM

Every organization uses a variety of tools to achieve institutional goals, including business plans, strategic plans, and long-range development plans. The goals of an organization's OHSP are as follows: to identify hazards in the workplace and determine the risk associated with them, to design the facility and management program to reduce risks associated with the hazards, and most importantly, to communicate hazard identification, risk assessment, and appropriate safety measures to all employees. An OHSP integrates the efforts of management, administration, employees, and health care professionals in an active, evolving program that promotes a culture of safety in the workplace.

The challenge of providing a safe work environment is best met with the development of an OHSP that provides a foundation for a culture of safety and makes worker safety a central mission for all employees of an institution. Inclusion of safety in the development of a new institution is generally easier than integration of safety into long-established programs. There is always the concern that worker safety and the attendant OHSP expenses will have adverse effects on finances and process efficiency. Although economics will have an impact on any animal care and use program, cost alone must not dictate the scope or relevance of the OHSP implemented at an institution. The simple trade-off is that employee welfare and reduction in the loss of work time due to workplace injury will improve employee satisfaction and performance. It is important for staff to know that management is concerned about their welfare. For both new and long-established institutions, there is value in having a reference document, such as this volume, that provides a ready source of information for creating an OHSP. The intent of this report is to provide the proper tools to identify and manage human health hazards associated with nonhuman-primate research.

The National Research Council developed a document on OHS for animal research facilities (NRC 1997), which serves as a guide for management of an OHSP and provides a foundation for developing a program if none exists. The present report is aimed at developing an OHSP at nonhuman-primate facilities or facilities that use nonhuman-primate blood or tissue and is not intended to duplicate the scope or content of the previous document. Instead, its goal is to complement that publication and expand on topics that are particularly relevant or peculiar to animal programs that involve nonhuman-primate species. The Committee on Occupational Health and Safety in the Care and Use of Nonhuman Primates has also attempted to address the meaning and implications of uncertainty in risk management.

OVERVIEW

This report is organized to address the various considerations in the development of an OHSP for nonhuman-primate facilities. The report follows a logical progression to facilitate its use in OHSP development, but each chapter can be used as an individual reference document.

The second chapter provides a background for the use of nonhuman primates in research, education, and testing and discusses the goal of this report—promoting OHS in nonhuman-primate facilities. To ensure that OHS issues bridge academic or departmental divisions, this chapter also highlights the importance of the Institutional Animal Care and Use Committee (IACUC) in implementing and monitoring the efficacy of the OHSP. Finally, this chapter addresses the taxonomic and behavioral diversity of the primate order. Understanding differences in behavior between species and individual animals can assist personnel in predicting an animal's actions and identifying potential hazards associated with them.

The third chapter addresses potential zoonotic hazards that may be encountered in a nonhuman-primate facility or at a facility that uses nonhuman-primate blood or tissue. The chapter is organized by type of disease agent—viral, bacterial, protozoan parasitic, metazoan parasitic and other agents. For each agent, there is a description of the disease profile in nonhuman primates, mode of transmission, incubation period and clinical signs, and diagnosis and prevention.

Chapter 4 identifies and describes the noninfectious hazards that are specific to facilities and research involving nonhuman primates. This chapter emphasizes that the identification of noninfectious hazards must involve a qualified health and safety professional who is trained in ergonomic hazards. This chapter also addresses potential hazards associated with allergies to nonhuman primates, heat stress associated with the personal protective equipment used when interacting with nonhuman primates, volatile anesthetics, and disinfectants commonly used in nonhuman-primate areas.

The fifth chapter discusses risk assessment as a powerful tool that provides a rational framework for designing and managing OHSPs in institutions that care for nonhuman primates. The Committee examines risk assessment as a process of collecting and analyzing scientific data to describe OHS risks. It identifies four steps to successful risk assessment: hazard identification, dose response assessment, exposure assessment, and risk estimation and characterization (NRC 1983; Samet and Burke 1998).

Chapter 6 discusses the identification of pertinent OHS regulations and recommendations, the final step undertaken before a risk management strategy is developed. It notes how this aspect of the process can be challenging because of the multiple agencies or regulations that may be applicable in the same facility and identifies the most widely relevant US federal regulations. The committee also discusses oversight responsibility and identifies important reference material and organizations that provide additional assistance and frameworks for developing an OHSP.

Chapter 7 reviews the steps necessary to formulate and implement a course of action to manage hazards identified during the risk assessment process. This course of action or risk management is the core of an OHSP. The chapter is divided into nine key areas through which hazards can be effectively managed, and each area contains a checklist of considerations to address for large and small institutions. The nine areas are: administrative procedures, facility design and operation, exposure-control methods, education and training, occupational health, equipment performance, information management, emergency procedures, and program evaluation.

Chapter 8 identifies the need for institutions to implement an inclusive training program for all personnel who interact with nonhuman primates, and a policy of continuing education to ensure that employees are informed and knowledgeable about hazards in the workplace. This chapter addresses critical issues in training programs that are designed to provide effective hazard communication to all employees regardless of educational level. The importance of worker orientation, standard operating procedures (SOP), and training evaluation is emphasized.

The final chapter presents an overview of medical management of persons involved in a nonhuman-primate-related injury or exposure. Establishing a working relationship in advance with a health-care professional is critical to determine what is an exposure, what is appropriate emergency treatment, and what are the options for postexposure prophylaxis.

CONCLUSION

The following chapters provide the foundation for the development of an OHSP at institutions engaged in the care and use of nonhuman primates. The Committee encourages program managers to utilize other resources as well, in particular the 1997 NRC document *Occupational Health and Safety in the Care and Use of Research Animals* and the 1999 CDC *Biosafety in Microbiological and Biomedical Laboratories*. The actual development and implementation of an OHSP must be adapted to the individual needs and functions of an organization. One common theme is that effective communication of hazards, risks, and safety measures to all employees is a vital element in the success of any program.

2
Background and Context for Occupational Health and Safety in the Care and Use of Nonhuman Primates

INTRODUCTION

According to annual reports published by the US Department of Agriculture, the number of nonhuman primates used or intended for use in research has remained generally stable for the last decade at about 52,000 animals per year. Over 87% of the states, districts, and territories in the United States use primates in research. Primates are used in diverse projects, including research in infectious diseases, cancer, neuroscience, heart disease, nutrition, and reproduction; drug development and safety assessment; and behavioral studies (Sibal and Samson 2001). The steady and widespread use of nonhuman primates strongly suggests that they will continue to be important animal models for a number of human diseases.

Many research projects require physical proximity between people and nonhuman primates or their tissues, and institutions have an ethical responsibility to provide for the health and safety of people exposed to the hazards resulting from that proximity. This responsibility is particularly important at institutions that use primate species that pose known and significant infectious hazards, such as the macaques and chimpanzees. Although economics will have a role in any animal care and use program, cost alone must not dictate the scope or relevance of the OHSP implemented at an institution.

A successful OHSP has seven basic elements: knowing the hazards, avoiding and preventing exposures to the hazards, providing effective training and reinforcement, promulgating and enforcing sound rules and guidelines, ensuring consistency and reliability in occupational behavior, maintaining proper records (a database) and providing comprehensive medical surveillance and feedback, and developing key leadership and staffing based on a commitment to a safe workplace (NRC 1997).

The importance of an OHSP in any laboratory animal care and use program is highlighted in the *Guide for the Care and Use of Laboratory Animals* (NRC 1996). The *Guide* identifies the essential elements of an OHSP, although more specifically in the context of animal care and use: hazard identification and risk assessment; personnel training; use of personal protective equipment; facilities, procedures, and monitoring; medical evaluation and preventive medicine; and addressing animal experimentation that involves hazards.

One of the most commonly identified deficiencies in animal care and use programs evaluated by the Association for Assessment and Accreditation of Laboratory Animal Care International (AAALAC International) is in the OHSP (DeLong and others 2001). Overarching concerns identified by AAALAC International were a failure to base the OHSP on hazard identification and risk assessment and a failure to include in the program the general hazards of working with animals (as opposed to experimental hazards themselves, which are usually taken into account); inadequate training on such OHS topics as zoonoses and allergies; inadequate inclusion in the OHSP of all personnel potentially at risk, such as students, the nonaffiliated member of the IACUC, and visiting scientists; and inadequate linkage between the IACUC and institutional safety personnel. AAALAC International's expectations for a sound OHSP are that the individual components of the program are appropriate for the institution and that the components work together effectively. Thus, there must be sound implementation strategies and effective coordination of program components and personnel.

INTENT OF THIS REPORT

The National Research Council report *Occupational Health and Safety in the Care and Use of Laboratory Animals* (1997) provides an excellent overview of OHS issues. However, in light of AAALAC International's findings and the potentially serious risk that working with nonhuman primates can pose to workers, the present report is intended to serve as a tool for developing and improving OHSPs that must address the particular hazards posed by nonhuman primates. The intent of this report is to provide specific information to safeguard the health of people working with nonhuman primates; however, an additional benefit is the protection of animal health. This report will serve as a resource for health care providers, institutional officials who are responsible for the OHSP, and the various personnel working with nonhuman primates or their tissues. References provided throughout the report will assist people who seek more detailed information on particular aspects of an OHSP.

The information presented here is intended to guide the implementation of an OHSP that must address the use of nonhuman primates and that is appropriate for the size and function of the institution. Any analysis of the risks posed by nonhuman primates to people must take into account the specific animal use (for example, how direct is the contact with animals or animal tissues), the nature of nonhuman primates being used (including species, age, sex, and previous experience of the animals), the qualifications of the people working with the animals or animal tissues, as well as support staff who might come into contact with the animals (e.g., maintenance or housekeeping personnel) or animal-associated equipment (e.g., cage-wash staff). Practices, procedures, and attitudes in place at an institution are essential components of an OHSP. Because the hazards associated with the use of nonhuman primates will be specific for each institution, no single approach to addressing the risks can be offered in this report. Instead, general guidance is provided on the types of hazards encountered in different settings with various primates and on how to avoid those hazards and manage exposures.

IMPLEMENTING THE OCCUPATIONAL HEALTH AND SAFETY PROGRAM

Responsibility, Authority, and Accountability

A successful OHSP begins with strong administrative support. The senior official at the institution must understand the health and safety issues related to working with nonhuman primates, support the development and implementation of policies to safeguard workers, communicate the importance of OHSP participation to them, ensure that suitable funding and other resources are available to implement and maintain the program, designate appropriate staff to serve on the design and implementation team, and identify the individual or office that will manage the OHSP.

The design and implementation team should comprise persons who have expertise in providing occupational health care and who might be exposed to hazards in the workplace, depending on their role in the institution. The team should include representatives of animal care and use staff, research staff, environmental health and safety staff, occupational health and medical staff; administration and management (NRC 1997) assisting with the design and implementation of the OHSP should be qualified to do so through training or experience with relevant hazards.

Oversight

A successful OHSP relies on the identification of health and safety concerns at all levels of the institution. Animal care and use staff typically have the most frequent and direct contact with nonhuman primates and therefore typically have a practical approach to identifying and reducing potential hazards. The IACUC, through its animal study proposal review process and semiannual facility inspections, has the opportunity to investigate possible risks associated with a proposal, to seek ways to reduce the risk, and to assess the animal facility and animal procedure areas for potential hazards that should be avoided or minimized.

The Environmental Health and Safety (EHS) staff, whether they are employees of the institution or contract staff, may be involved in radiation and chemical safety, waste management, and monitoring of the OHSP. EHS personnel often provide input into the development of work practices and should maintain a liaison with the IACUC for review of animal-study proposals involving hazardous agents. In institutions where nonhuman primates are used, the EHS staff should have appropriate knowledge of all associated infectious, chemical, and physical hazards and ergonomic concerns, along with the importance of protection of the animals from human disease. Persons assigned these responsibilities can be invaluable in the implementation of a program modification. Information gathering is particularly important when research programs include novel infectious agents or viral vectors on which little reference information is available. The senior official, through reports provided by the research directors and staff, the IACUC, the EHS and OHS directors and staff, and the NHP director and animal care staff provides the highest level of oversight by ensuring that problems identified by others in the institution are resolved (Figure 2-1).

The OHSP is also reviewed by external oversight organizations, such as OSHA and corresponding state agencies, the National Institutes of Health Office of Laboratory Animal Welfare (for institutions receiving Public Health Service [PHS] funding), and AAALAC International (for institutions seeking or renewing accreditation). Those organizations can take different actions when an institution is not in compliance with their standards, such as imposition of fines, withdrawal of grant funds, and revocation of accreditation, respectively.

Hazard Identification and Risk Assessment

Risk assessment is the basis for designing and managing occupational health and safety programs to reduce workplace risks to an acceptable level. Risk assessment measures the likelihood of adverse health effects resulting from occupational injuries or exposures and is discussed further

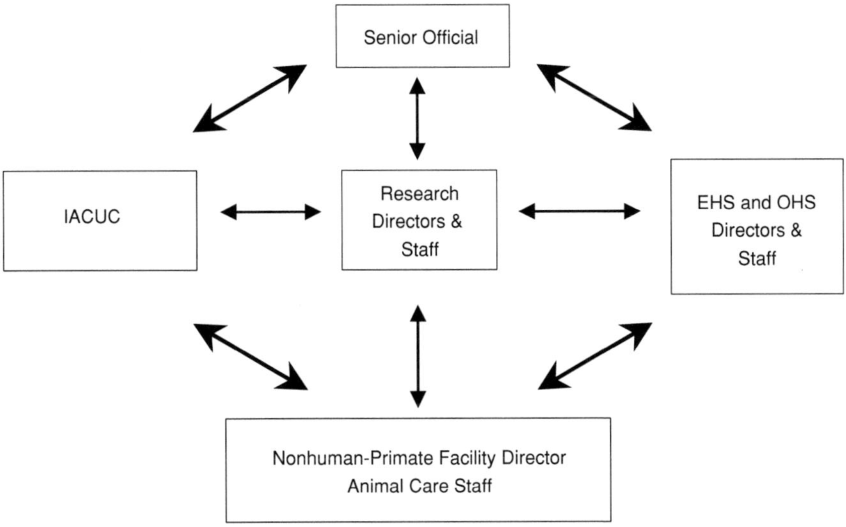

FIGURE 2-1 Relationships among institutional administrators and units responsible for occupational health and safety in the nonhuman primate facility.

in Chapter 5. An important aspect of risk assessment is hazard identification. A hazard is a source of risk, such as a substance or action that can cause harm. Hazards can be inherent to working in an animal facility or result from a specific research project. Everyone in the facility should be considered responsible for identifying hazards as part of his/her duties to ensure a culture of safety in the institution. Hazards can be categorized as biologic, chemical, or physical, and they can be introduced by dermal or mucous membrane contact or through respiratory, oral, or auditory routes. Occupational hazards associated with nonhuman-primate facilities are principally those related to physical injury and infectious disease (viral, bacterial, and parasitic). The process of identifying hazards should be continuous as new equipment or materials, new species, and new research are introduced into the institution. Experience in and qualifications for working with one species of primate cannot necessarily be translated to a different species, even if the species belong to the same genus. Therefore, hazard identification must include knowledge of the biology and behavior of the primate species in question. Table 2-1 shows some of the key factors involved in identifying hazards and then conducting a risk assessment. Risk assessment should be used to manage identified hazards, avoid or minimize potential exposures, and guide treatment if an exposure occurs.

TABLE 2-1 Factors Involved in Risk Assessment

Known Hazards:
- Experimental conditions
- History of health concerns in facility/program

Unknown Hazards:
- Employment outside primary workplace
- Hobbies, circumstances at home that may predispose for risk at primary workplace (e.g., sports, pets)

Work Assignment:
- Duration of study
- Frequency of exposure
- Protection afforded through regulatory (e.g., CDC) requirements

Species:
- Of animal
- Potential for zoonotic disease
- Potential for injury
- Of experimental agent
- Specific agent properties
- Intensity of exposure

Protection Afforded Through:
- Facility engineering
- Required personal protective equipment
- Required biosafety equipment
- Health status of worker
- Experience of worker

SOURCE: Adapted from AAALAC International, see www.aaalac.org.

HAZARDS ASSOCIATED WITH NONHUMAN-PRIMATE BEHAVIOR

Personnel working with nonhuman primates should receive basic training in nonhuman-primate behavior. Understanding the behavior of primates assists personnel in predicting the animals' actions and identifying potential hazards. Such an understanding includes a basic knowledge of the animals' anatomy and perceptual capabilities; this information indicates the physical capabilities of the animals (leaping or spitting long distances, running rapidly, exercising manual dexterity, and so forth) and what it perceives in its environment (such as depth, color, and smells). Knowledge of the animals' cognitive abilities aids colony managers, scientists, and veterinarians in selecting housing or testing equipment that is appropriate and enhances personnel safety and in determining what kind of training program can be used with the animals. Knowledge of animal behavior is an important factor in the safe handling of large animals (Grandin 1999) such as nonhuman primates. This background information can be overlaid with detailed knowledge regarding individual animals, such as an animal's previous use (and familiarity with people and procedures) and level of aggressiveness or other behavior, which can be relevant to personnel safety.

Standard operating procedures (SOPs) or institutional policies are critical to safeguarding people at risk. Such documents may restrict entry to the animal facility, require that visitors be older than some minimal age (e.g., 18 years), or require particular health checks of visitors (for example, tuberculosis testing and measles vaccination) to protect both the animals and the people. SOPs may specify a variety of procedures for husbandry or research that maximize animal well-being and minimize potential risk to the individual. SOPs and institutional policies should be customized to the facility and kept current by periodic review (for example, by the IACUC).

Occupational hazards associated with nonhuman primates can occur in zoologic parks, in research and testing environments, and in teaching or breeding programs. Thus, several categories of people, perhaps with different training in primate work, are potentially at risk. People at risk include employees; students, visiting scientists, and other trainees; contract workers; and visitors or guests to the facility. The categories of people at risk vary not only in expertise, but also in physical proximity to the animals, the type of use of the animals (for example, the animal might be conscious or unconscious), and the species and number of animals being handled (i.e., the risk posed by working with some species can increase with the number of animals).

Training programs should be made available to individuals at risk to provide information on potential hazards and avoidance of risks. Depending on the primate species and the target audience, the training program should address zoonotic diseases of primates, ergonomic hazards, methods of reporting injuries, first aid, follow-up health care, and principles of primate behavior. One-on-one training by supervisors is critical to tailor the information conveyed so that it is specific to the task and species to which the trainee will be exposed. A combination of didactic and hands-on training may optimize personnel safety.

The order Primata is diverse, comprising, in addition to humans, more than 200 species of prosimians, New World monkeys, Old World monkeys, and apes. That diversity is reflected in the wide variety of habitats occupied by nonhuman primates and by their anatomic, physiologic, and behavioral differences. Of the numerous nonhuman species, relatively few are used in research, although a larger number are maintained in zoological parks. Some of the behavioral considerations in handling those species most prevalent in research are discussed below. A detailed discussion of this topic is included in *The Psychological Well-Being of Nonhuman Primates* (NRC 1998).

Prosimians

The prosimians or lower primates (suborder Strepsirhini) include tarsiers, lemurs, sifakas, indris, aye-ayes, lorises, pottos, and galagos (Nowak 1999). Lemurs and lorises are often seen in zoos. The prosimians are a diverse group, whose members range from 12 cm and less than 100 g (the mouse lemur) to 90 cm and 10 kg (the indri). Prosimians are considered the primates most removed taxonomically from humans, and there has been no published report of disease transmission from prosimians to humans (NRC 1998). In general, prosimians are not aggressive, although some species actively resist restraint by kicking and biting. The bite of a slow loris may be of particular concern because of its poisonous mix of saliva and glandular secretions (Alterman, 1995). Some prosimians are inquisitive and may leap onto people to investigate them more closely. Personnel entering these animals' quarters should be prepared for such behavior.

New World Monkeys

The New World monkeys are found in Central America and South America. They include marmosets and tamarins, which are collectively known as callitrichids (Callitrichidae), as well as squirrel monkeys, owl monkeys, titi monkeys, capuchin monkeys, spider monkeys, howler monkeys, woolly monkeys, sakis, and uacaris, which are collectively known as cebids (Cebidae). The common marmoset (*Callithrix jacchus*) is used in biomedical research and in zoological exhibits; other callitrichids are typically only maintained in zoos.

The callitrichids are distinct from the cebids by having claws on most digits. They are also principally arboreal, descending to the ground only occasionally. Callitrichids are highly territorial, and their territoriality can occasionally be directed at personnel coming close to their housing enclosure. Marmosets and tamarins have well-developed visual, olfactory, and auditory perception and long memories, so they are able to recognize individual humans and can develop strong likes and dislikes of them (NRC 1998). Threatening behavior may be detected in some species of callitrichids in an arching of the back and concomitant stiff-legged walk, presentation of the testes, or a chest display (achieved by standing bipedally and turning the elbow out) (Hershkovitz 1975). Occasionally, the bipedal stance will be accompanied by swaying from side to side. In addition, the entire pelage may be piloerected in a threat display. Hand-raised animals can become very aggressive toward people when they reach puberty. Their territorial and occasionally aggressive behavior, in combination with sharp claws and procumbent incisors (marmosets) or long canine teeth (tamarins), poses a risk of bite wounds to personnel.

Of the cebids, squirrel monkeys and less often capuchins, spider monkeys, and owl monkeys are used in research. These and the other cebids can also be found in zoological exhibits. In general, New World monkeys are not aggressive toward humans and do not respond aggressively to direct eye contact, as Old World monkeys do (NRC 1998). Like prosimians, however, they resist restraint vigorously. Capuchins have excellent manual dexterity and manipulative abilities; these characteristics have resulted in their occasionally unlocking their enclosures and getting free in an animal holding room. Staff should exercise care in capturing these animals because they might bite in self-defense or when frightened. Similarly, a cebid attempting to climb on a person might bite the person if pushed away or frightened (NRC 1998). Occasionally, a cebid will defend a person it likes from other people.

Old World Monkeys

The Old World monkeys are native to Africa and Asia, although introduced populations exist throughout the world. The family of cercopithecids (Cercopithicidae) comprises two subfamilies, the cercopithecines (Cercopithecinae) and the colobines (Colobinae). The macaques, baboons, drills/mandrills, geladas, mangabeys, guenons, talapoins, African green monkeys (a.k.a. grivets or vervets), patas monkeys (military monkeys, hussar monkeys, and mustached monkeys), and Allen's swamp monkeys are cercopithecines. The colobines include colobus monkeys, langurs, Chinese golden monkeys, and proboscis monkeys. Of the cercopithecids, the rhesus monkey and cynomolgus monkey are the most commonly used in research; baboons, African green monkeys, and a few other species of macaques are also used. Most of the cercopithecids and colobines can be found in zoological parks.

Many species of Old World monkeys have well-developed cognitive abilities, although most studies of cognition have been done in macaques. The cercopithecines are also strong for their body size and are skilled at manipulating objects. Studies have demonstrated macaques' puzzle-solving ability, even in the absence of a food reward (Washburn and Rumbaugh 1992). The combination of cognition, strength, and propensity to manipulate objects can result in animals' escaping from their enclosures, thereby posing a risk to personnel. Macaques have well-developed visual capabilities (Bayne and Davis 1983; DeValois and Jacobs 1971; Leary and others 1985), and visual signals (such as coloration, facial expressions, and body posture) are important in conspecific communication. A lack of knowledge or understanding of the visual signals on the part of people working with macaques can lead to the inadvertent communication of mild threats, such as through direct eye contact and jerky arm movements, and the primates might respond to these perceived threats aggressively. Because the cercopithecines are generally social animals, any action by humans that is perceived as threatening to an infant (such as removing the infant from a group) can also elicit aggression. Basic training of key personnel in primate behavior can decrease the primates' aggressive actions in response to human behavior. Risk can be reduced in some instances by training the animals (for example, macaques) to cooperate in specific activities, such as cage transfer, venipuncture, and vaginal swab sampling. This can minimize the handling of conscious animals and reduce the need for chemical immobilization.

Apes

The superfamily of apes or hominoides (Hominoidea) consists of the hylobatids (Hylobatidae), or lesser apes, and the hominids (Hominidae), which includes great apes and humans. Some taxonomists place the orangutan in its own family (Pongidae). Gibbons and siamangs make up the lesser apes. The great apes comprise chimpanzees, bonobos, gorillas, and orangutans. Both lesser and great apes are frequently maintained in zoological parks; the chimpanzee is the most common ape used in research.

The lesser apes exhibit specialized locomotion known as brachiation (arm-swinging) aided by their long arms and elongated hands and fingers. Those features also result in a long reaching grasp through enclosure barriers to grab at unwary people who are too close. Gibbons are capable of rapid movement, and they may bite (NRC 1998).

The great apes are extremely strong, including in their hands. Chimpanzees and gorillas will use objects in their environment in their charge displays, occasionally throwing the objects at specific targets. Chimpanzees and orangutans will also spit saliva or water and throw feces at nearby persons with great accuracy. A response to this behavior from a person encourages it to the point of becoming routine in the animal's behavioral profile. Young chimpanzees can pinch and bruise people with their rough play. Great apes have a highly developed cognitive level, with excellent skills in complex learning and tool-using. Some species are even capable of recognizing themselves in mirrors and in televised images (Gallup 1977, 1982; Lambeth and Bloomsmith 1992; Menzel and Lawson 1985). Great apes do not often trust unfamiliar persons and can even be devious in their relationships with people (NRC 1998), which can place those individuals at additional risk. Chimpanzees have long memories, and their aggressive actions can be unpredictable and appear premeditated. Enclosures should be designed to prevent great apes from reaching out and grabbing people; nonhuman primates have been reported to grab neckties, loose-fitting laboratory coats, or long hair (NRC 1997). Because of their cognitive skills, great apes can be trained by using a food reward for cooperation in such activities as venipuncture and injections. This training can reduce the need to sedate the animals while minimizing risks to personnel.

RISKS AND RISK REDUCTION ASSOCIATED WITH ENVIRONMENTAL ENRICHMENT

Ethical reasons and federal regulatory requirements mandate that an environmental enrichment program be provided to captive nonhuman primates to improve their well-being. The *Guide* (NRC 1996) suggests that this enrichment be provided vis-à-vis a behavioral management program that comprises three principal elements: the structural environment, the social environment, and activity. An appropriately designed behavioral management program can be a useful tool to reduce risks associated with working with nonhuman primates. However, each of the components of the behavioral management program also poses its own OHS challenges for personnel.

A key way in which a behavioral management program can improve worker safety is by reducing atypical behavior expressed by nonhuman primates. Animals that are exhibiting behavioral pathology resulting from their captive conditions can be unpredictable and excessively aggressive, thereby increasing the risk of injury to workers. Efforts to maintain animals in a state of well-being so that they express species-typical, and thus more predictable, behaviors will improve worker safety.

In addition, when the animals associate the presence of personnel with positive experiences, such as the provision of food treats or cognitive activities, their behavior toward staff will more likely be affiliative rather than aggressive. As described previously, training animals to participate in routine procedures, as a part of the behavioral management program, reduces the risk of injury to workers because the animals will be cooperating in the activity rather than resisting.

Care must be exercised when personnel are forming social pairs or groups of primates for enrichment purposes because the animals will form dominance hierarchies through acts of aggression and submission that can inadvertently involve staff members. Published reports have described methods of forming pairs or groups to maximize the safety of animals (Bernstein 1991; Fritz 1994; Reinhardt 1988, 1990, 1991). Those methods will also protect the personnel involved.

Many enrichment techniques involve attaching items (such as foraging devices and toys) to the front of a cage or placing items inside the cage, so there is the potential for animals to grab, bite, or scratch personnel as they provide upkeep of these items. In general, enrichment items kept inside the cage should be handled only when the cage is empty (for example, during cage-change procedures). Good judgment must be used when considering the upkeep of items attached to the front of an animal's enclosure; the practice might be safe with some enrichment items and for some species of nonhuman primates but not others. Consideration should also be given to the risks associated with handling and transport of the enrichment devices themselves. These devices can become contaminated with saliva, urine and feces and can be an infectious hazard to any personnel that contact them, including staff that may wash the devices. Further, it should be noted that microbial growth can persist on enrichment devices after sanitation in a commercial cage washer (Bayne and others 1993).

Because the provision of environmental enrichment is a required aspect of captive-primate husbandry, staff with a variety of expertise may be involved in the behavioral management program. Institutions should ensure that staff receive training in the safe implementation of enrichment that is specific to the species held, the type of caging used, and the methods of enrichment being implemented.

3
Identifying Infectious Hazards Associated with the Use of Nonhuman Primates in Research

Many pathogenic organisms that naturally infect nonhuman primates are communicable to humans, and several human pathogenic organisms are communicable to nonhuman primates and can be retransmitted back to humans. Because humans and nonhuman primates have a close phylogenetic relationship, the risk of transmission of pathogenic organisms with nonhuman primates is greater than with any other group of laboratory animals used in biomedical research. This potential risk increases the importance of identifying infectious hazards for persons working with nonhuman primates or their blood or tissue.

Pathogenic organisms can be acquired by exposure to blood or body fluids by any route including needle inoculation, animal bites and scratches, splashes, accidental ingestion, mucous membrane contamination, contaminated caging and equipment, or even infectious aerosols. Furthermore, some of these organisms may be significantly more pathogenic in species that are not naturally exposed (for example, B virus causes mild, self-limiting lesions in macaques, but is highly pathogenic and often fatal for humans and other nonhuman-primate species such as marmosets and capuchins). The invaluable use of these animals as models of human infectious disease compounds these concerns; their care and use after inoculation with hazardous pathogens can entail substantial hazards. The tendency for zoonotic agents to cause asymptomatic infections in their natural host species raises additional considerations, as does the sharing of nonhuman-primate blood and tissues among laboratories where the procedures and safeguards in place might not be aligned with the etiologic hazards in mind.

The diversity of nonhuman-primate species in research and the lack of comprehensive knowledge regarding the biology and epidemiology of each agent make the list of infectious hazards summarized in this chapter necessarily incomplete. However, the agents described here span the taxonomic groups of pathogens described to date, including those with the most clearly documented importance for laboratory primate research colonies. New findings should be appended to these listings, and the adequacy of safety programs should be reviewed accordingly. Criteria for inclusion in this chapter were the presence of published case reports of occupational exposures, the existence of population-based surveys in research settings or native habitats, and the biologic plausibility of accidental human exposures. The plausibility is clearly less for agents that require intermediate hosts, arthropod vectors, or environmental incubation, but accidental inoculation via a penetrating injury nonetheless warrants inclusion even for some of those agents. For ease of reference, agents are described at the host genus or species level, as appropriate. Some examples of infectious hazards introduced experimentally into nonhuman primates are considered, and assessments of their potential for human exposure should be re-examined in actual institutional contexts. However, a comprehensive review of the hazards associated with experimentally-induced infections in NHP is beyond the scope of this report.

Most agents likely to be encountered in common species in research use are listed in Table 3-1. Some significant taxonomic groups of nonhuman primates used less commonly in contemporary scientific studies (such as marmosets, owl monkeys, and mangabeys) have been largely excluded from the table because of insufficient descriptions of their potential role in infectious hazards, so this chapter should not be considered exhaustive. Furthermore, some nonspecific agents (such as dermatophytes and rabies) should be considered potential hazards from any species of nonhuman primate. Finally, the possibility of zoonotic disease transmission arising from xenotransplantation of nonhuman-primate tissues to humans raises a variety of additional concerns (Michaels 1998) that are beyond the scope of this work.

The information on infectious agents to which humans may be exposed through contact with nonhuman primates is organized into major sections of viral diseases, bacterial diseases, protozoan parasite diseases, metazoan parasite diseases and other agents of potential importance within the context of contemporary animal care and use programs. Information relevant to each agent is presented in four categories: disease profile in nonhuman primates, mode of transmission, incubation period and clinical signs, and diagnosis and prevention. It should be clarified that not all of the agents listed have been recognized as causes of any illness or other untoward effect in human beings to date. However, given the slow rate of disease progression for some recognized human pathogens, appreciation for the potential of genetic recombination and multiple causation of disease, and the relatively short time since their discovery, the goals of an OHSP should be to prevent exposure to these agents in the work environment regardless of their suspected importance. Those responsible for the OSHP should take into account the potential for exposure to all of these agents in occupational settings that involve nonhuman primates or their tissues. Detailed considerations for the design and implementation of an OHSP relative to infectious and non-infectious hazards are presented in Chapter 7 of this report.

VIRAL DISEASES

Several virus classes are chronic or latent infections of a given species of nonhuman primate and are discussed below or listed in Table 3-1. These are likely to be present in all species of nonhuman primate although they are not necessarily described for a given species. They include herpesviruses, foamy viruses, and papovaviruses. Some species also have their own hepatitis A and B viruses as well as chronic blood-borne flaviviruses and speculatively could participate in transmission of agents of nonhuman primate or human origin.

The lentivirus taxon is particularly important because of its phylogenetic relation to human acquired immune deficiency syndrome (AIDS). It is likely that human immunodeficiency virus 1 (HIV1) originated from a chimpanzee lentivirus (Gao and others 1999) and that HIV2 derived from a sooty mangeby lentivirus (Hirsch and others 1989). Only African nonhuman-primate species have been definitively shown to be chronically infected with their own specific lentivirus. Macaques are an important consideration because they are often used as experimental models of AIDS after inoculation of lentiviruses derived from other nonhuman primates. They may be chronically viremic and a source of blood-borne infection to humans or after immunodeficiency develops may be amplifiers of opportunistic infections.

Examples of the most significant of these types of viral infection are discussed below.

B Virus

Disease Profile in Nonhuman Primates

B virus, also known as *Herpesvirus simiae* and Cercopithecine herpesvirus 1, is an alphaherpesvirus that occurs naturally only in macaques (Holmes and others 1995; Weigler 1992). The pathogenesis of B virus in

TABLE 3-1 Infectious Hazards from Nonhuman Primates

	Macaques	Baboons	Guenons	Squirrel Monkeys	Chimpanzees
Viruses	B virus Foamy virus Simian retrovirus (Type D) SV40 SIV Pox viruses Yellow fever Dengue Ebola	Foamy virus Pox viruses Yellow fever Dengue	Foamy virus SIV Pox viruses Yellow fever Dengue	Dengue Yellow fever	Foamy virus SIV Hepatitis B Molluscum contagiosum Hepatitis A Pox viruses Yellow fever Dengue Ebola
Bacteria	*Burkholdria pseudomallei* *Campylobacter* spp. *Mycobacterium tuberculosis* *Mycobacterium bovis* *Mycobacterium leprae* (also known in mangabeys) *Leptospira* spp. *Salmonella* spp. *Shigella* spp. *Yersinia pseudotuberculosis* *Yersinia enterocolitica*	*Campylobacter* spp. *Leptospira* spp. *Mycobacterium tuberculosis* *Mycobacterium bovis* *Salmonella* spp. *Shigella* spp. *Yersinia pseudotuberculosis* *Yersinia enterocolitica*	*Campylobacter* spp. *Leptospira* spp. *Mycobacterium tuberculosis* *Mycobacterium bovis* *Salmonella* spp. *Shigella* spp. *Yersinia pseudotuberculosis* *Yersinia enterocolitica*	*Campylobacter* spp. *Leptospira* spp. *Mycobacterium tuberculosis* *Mycobacterium bovis* *Salmonella* spp. *Shigella* spp. *Yersinia pseudotuberculosis* *Yersinia enterocolitica*	*Burkholdria pseudomallei* *Campylobacter* spp. *Mycobacterium tuberculosis* *Mycobacterium bovis* *Mycobacterium leprae* *Leptospira* spp. *Salmonella* spp. *Shigella* spp. *Yersinia pseudotuberculosis* *Yersinia enterocolitica*

Metazoan Parasites	*Hymenolepis nana* *Oesophagostomum* spp. *Strongyloides* spp. *Trichuris* spp. *Enterobius vermicularis*	*Hymenolepis nana* *Oesophagostomum* spp. *Strongyloides* spp. *Trichuris* spp.	*Oesophagostomum* spp. *Strongyloides* spp. *Trichuris* spp.	*Hymenolepis nana* *Trichuris trichuria*	*Hymenolepis nana* *Oesophagostomum* spp. *Strongyloides* spp. *Trichuris* spp. *Enterobius vermicularis*
Protozoan Parasites	*Balantidium coli* *Cryptosporidium* spp. *Entamoeba histolytica* *Giardia intestinalis* *Plasmodium* spp.	*Balantidium coli* *Cryptosporidium* spp. *Entamoeba histolytica* *Giardia intestinalis* *Plasmodium* spp.	*Balantidium coli* *Cryptosporidium* spp. *Entamoeba histolytica* *Giardia intestinalis* *Plasmodium* spp.	*Balantidium coli* *Cryptosporidium* spp. *Entamoeba histolytica* *Giardia intestinalis* *Plasmodium* spp. *Trypanosoma cruzi*	*Balantidium coli* *Cryptosporidium* spp. *Entamoeba histolytica* *Giardia intestinalis* *Plasmodium* spp.

macaques resembles that of herpes simplex viruses 1 and 2 in humans, with primary infections followed by lifelong latency, mostly in trigeminal and lumbosacral sensory ganglia, and recrudescent episodes of virus shedding on one or more occasions from tissues around the original site of virus exposure. Despite the shedding of infectious virus at those times, most macaques with B virus infections are asymptomatic or experience only mild, self-limiting, localized lesions that are often not outwardly apparent. When present, vesicles, progressing to ulcers that heal without scarring after 10-14 days, are generally noted on oral-facial or genital mucous membranes and mucocutaneous borders, often including the conjunctivae. Disseminated disease in macaques involving widespread hemorrhagic necrosis of the liver, lung, brain, and lymphoid organs occurs rarely (McClure and others 1973). Natural transmission of the agent between macaques occurs principally via biting and scratching but also via sexual activity in postpubertal animals (Weigler and others 1993, 1995). Prevalence can reach 90% or more in group-housed breeding colonies (Weigler and others 1990). Single- or pair-housing configurations tend to have lower prevalence (Weir and others 1993), but infection status and likelihood of virus shedding are unpredictable, so it is prudent to consider all macaques to be latently or actively infected with this agent (Ward and Hilliard 2002). B virus is also known to be highly pathogenic in other species of nonhuman primates including colobus monkeys, patas monkeys, DeBrazza's monkeys, capuchins, and marmosets in contact with infected macaques or after experimental inoculation with the agent (Loomis and others 1981; Thompson and others 2000; Weigler 1992; Wilson and others 1990).

Mode of Transmission

B virus exposures in humans have resulted from animal bites and scratches, splashes, needle stick injuries (although this virus is not considered a bloodborne pathogen), and other contact of mucous membranes or broken skin with infected body fluids from macaques or with wet, unfixed tissues or primary cell culture tissue material. Contaminated husbandry and research equipment can potentially spread B virus, although its viability is not expected to be prolonged (less than 24 hours in most cases), especially when subject to drying or sunlight. However, B virus may survive for longer periods when protected from environmental exposure in certain laboratory settings (Hilliard and Henkel 1998). Severity of injury has not correlated with likelihood of B virus infection, and several human cases have been noted without clearly recognized exposure incidents.

Incubation Period and Clinical Signs

The incubation period prior to onset of clinical signs in humans is variable; typically it is about 2-3 weeks. Some cases were apparent after a few days, and others have taken several weeks or more to manifest after the likely exposure. Postexposure wound cleansing and antiviral prophylaxis can greatly alter these patterns.

Numerous early-stage symptoms have been reported, including unexplained febrile disease (fever, chills, nausea, vomiting, and dizziness) and persistent headache. On some occasions, fluid-filled vesicles have formed near skin wounds sustained from macaque bites or scratches and have been followed by localized paresthesia. Mistaken early diagnoses have included influenza or sinusitis. Progression of disease may have other symptoms attributable to central nervous system infection, such as ascending encephalomyelitis, diplopia, seizures, and respiratory failure due to virus-associated tissue destruction generally localized to the brainstem (Whitley and Hilliard 2001).

B virus infection in humans has been documented on at least 50 occasions and has led to at least 23 deaths (Palmer 1987; Cohen and others 2002). When exposure is not evaluated promptly and there is no specific antiviral therapy, case fatality rate exceeds 80% (Hilliard, personal observations; Palmer 1987); thus B virus is the most significant infectious occupational health hazard in the conduct of nonhuman-primate research. Nonetheless, human cases of B virus are extremely rare despite its high prevalence in the host species and given the large numbers of macaques used in research for many decades.

As a cautionary note, healthcare workers should remember that all primate alphaherpesviruses studied to date are capable of establishing latency and therefore have the potential to reactivate. Given the existence of at least 6 individuals with persistent high titers of B virus antibodies and previous histories of clinically diagnosed B virus infections, the biomedical community can effectively remain vigilant to the possibility of reactivated B virus infections. Additionally, severe morbidity has been recognized clinically in at least 2 antibody positive individuals who reportedly had no contact with macaque monkeys for more than a decade prior to clinical presentation. Together, these data underscore the importance of patient follow-up to accumulate objective data in the face of existing knowledge of alphaherpesvirus latency and reactivation.

Diagnosis and Prevention

Diagnosis of B virus exposure in humans is through serology, virus isolation, and polymerase chain reaction (PCR) assays in reference laboratories that are capable of detecting low-level infections and differentiating cross-reacting antibody responses due to herpes simplex viruses (Holmes and others 1995; Katz and others 1986; Scinicariello and others 1993). Evaluation of involved macaques, tissues, or research materials may be useful in these assessments.

The use of barrier methods of protection and safe-handling procedures in work with macaques is paramount to B virus prevention. Prompt and sufficient attention to disinfecting or flushing of body sites known or thought to be contaminated, followed by proper follow-up evaluation and care, as dictated by a medical professional knowledgeable in this condition, is essential.

Several colonies of rhesus macaques that have no detectable antibody are known to exist (Ward and Hilliard 2002; Ward and others 2000), which should ultimately result in fewer cases of this disease in the coming decades, at least from this species of macaque. However, due to the inaccuracy of some B virus tests, it should be assumed that all macaques, including those from SPF colonies, are infected (Ward and Hilliard, 2002). Therefore animals from SPF colonies that are involved in an exposure should be as systematically and thoroughly evaluated as animals of unknown status.

Simian Immunodeficiency Virus

Disease Profile in Nonhuman Primates

Several closely related lentiviruses, designated simian immunodeficiency viruses (SIVs), have been found as persistent nonpathogenic infections in their natural reservoir in various species of Old World nonhuman primates from geographically disparate areas, including mangabeys, guenons, mandrills, and chimpanzees (Brown 1997; Mansfield and King 1998; Santiago and others 2002). When unknowingly or deliberately inoculated into macaques, they often cause an AIDS-like syndrome. Their close nucleic acid sequence homology with human immunodeficiency viruses (HIV-1 and HIV-2) and the pathologic and clinical patterns that follow SIV infections in macaques have given them an extremely important role as experimental models of AIDS biology (Mansfield and King 1998). In experimentally infected macaques, the incubation period can vary from weeks to months, depending on the model. SIV-inoculated macaques can succumb to a chronic wasting illness and an array of opportunistic infections, including cytomegalovirus, *Pneumocystis carinii*, *Mycobacterium avium* complex, *Cryptosporidium* spp., *Toxoplasma gondii*, and *Candida albicans*. Animals with naturally occurring and experimental SIV infections and the associated primate tissues (including blood and blood products) constitute potential infectious hazards to personnel. Potential zoonotic hazards include lentiviruses, and in the case of immunosuppressive infections, opportunistic pathogens.

The natural mode of SIV transmission among nonhuman primates is poorly defined, but there is evidence of sexual transmission in some circumstances. Transmission from infected dams to their offspring has been demonstrated (Phillips-Conroy and others 1994), and all infections are considered to be lifelong (Mansfield and King 1998).

Mode of Transmission

To date, three individuals have been infected with SIV, these occupational exposures occurred through splashes of infectious material onto mucous membranes, contamination of open cuts or abrasions on the skin, and needle stick injuries (Essex 1994; Khabbaz and others 1994; Sotir and others 1997).

Incubation Period and Clinical Signs

The incubation period for human cases is undefined, as no clinical signs of disease have occurred in exposed persons.

Diagnosis and Prevention

Detection of SIV infection in exposed persons can be via serology, virus isolation, and PCR. Use of barrier methods of protection and safe-handling procedures is warranted in work with nonhuman primates. Experimental studies with SIV are typically conducted under animal biosafety level (ABSL) 2 or 2/3 conditions (that is, ABSL-2 facilities with ABSL-3 practices and procedures) (see Table 5-2). Postexposure prophylaxis regimens involving the use of antiretroviral agents, as used for HIV case management, have been published (CDC 2001c).

Simian Foamy Virus

Disease Profile in Nonhuman Primates

Spumaviruses (simian foamy viruses), with close phylogenetic relationships to human foamy virus isolates, have been shown capable of transmission to humans. The involved species of nonhuman primates have included macaques, baboons, guenons, and chimpanzees (Brooks and others 2002; Heneine and others 1998; Sandstrom and others 2000). Species-specific isolates have also been recovered from New World monkeys such as squirrel monkeys and spider monkeys (Meiering and Linial 2001) and could potentially represent additional sources for human infection. There is no evidence that these agents are pathogenic in nonhuman primates or humans. These viruses are found frequently in cell cultures prepared by harvesting tissues for cell growth and propagation, complicating research by their presence.

Mode of Transmission

The mode of transmission to humans is unknown but presumed to be via contaminated saliva and possibly via bites and invasive research or veterinary procedures involving the oral cavity and respiratory tract of infected animals.

Incubation Period and Clinical Signs

No illness has been described as a result of spumavirus infections of human or nonhuman-primate origin.

Diagnosis and Prevention

Serology, PCR, and virus-isolation assays are used for diagnosis. Barrier methods of protection and safe-handling procedures in work with nonhuman primates should minimize the likelihood of occupationally acquired infections.

Ebola/Marburg/Filoviruses

Disease Profile in Nonhuman Primates

Imported macaques were implicated in outbreaks of Ebola subtype Reston (Ebola-R) among macaques in US facilities beginning in 1989 (Brown 1997; CDC 1989, 1990a, 1990b, 1991, 1996; Dalgard and others 1992; Miranda and others 1999; Rollin and others 1999; Schou and Hansen 2000), and guenons brought from Uganda were the source of Marburg virus exposure among laboratory workers in Germany and Yugoslavia in 1967 (Brack 1987). Ebola outbreaks in wild chimpanzees have also been reported (Formenty and others 1999). Nonhuman primates are unlikely to be the reservoir of Ebola virus since experimental or natural infection is quickly fatal (Georges-Courbot and others 1997).

Disease presentation after infection with these filoviruses in nonhuman primates has varied with the virus strain and species involved. Lymphoid necrosis, hepatocellular necrosis, interstitial pneumonia, and rapidly progressing (less than 24 hours) fatal hemorrhagic disease have occurred in recent outbreaks among imported macaques in the United States.

Mode of Transmission

The mode of Ebola virus transmission to humans is thought to be mainly by droplets and body fluid fomites although the filoviruses form infectious aerosols. Transmission of Marburg virus between animals and humans has usually involved contact with infected tissues (NRC 1998).

Incubation Period and Clinical Signs

The incubation period for these agents in humans is unknown. There have been no clinical signs in association with small numbers of Ebola-R infections in humans. Incubation for Marburg virus is 4-16 days. Initial clinical signs include fever, myalgia, and headache, followed by nausea, vomiting, and diarrhea along with thrombocytopenia and leukopenia (NRC 1997).

Diagnosis and Prevention

Diagnosis is based on detecting virus in acute disease samples from humans or other primates (antigen-detection ELISA, RT-PCR, or virus isolation). Ebola-R virus has not been recovered from human patients.

Prevention is through use of quarantine facilities approved by the Centers for Disease Control and Prevention (CDC) and appropriate biosafety programs for imported macaques and African green monkeys undergoing quarantine, especially after receipt from endemic areas. CDC has provided specific guidelines on the requirements for testing and health reporting for these species held in quarantine (CDC 1990a). Transmission of Marburg virus to humans in 1967 led to the institution of preventive quarantine measures that have stopped any significant spread of imported filoviruses and would be expected to be effective against many of the other viruses discussed in this report (Formenty and others 1999; Miranda and others 1999). The efficacy of these measures is based on the propositions that filoviruses result in significant overt disease in nonhuman primates held during quarantine, that the agents are not chronic or latent and recrudescent, that disease is properly evaluated when it occurs, and that there is oversight by persons whose primary concern is public health. Recent experience in this regard is illustrated in Box 3-1.

BOX 3-1
Some Lessons from the Ebola (Subtype Reston) Episode

The most recent outbreak of disease in nonhuman primates followed an introduction of a filovirus into a U.S. quarantine facility in 1989-1990. It led to several observations relevant to the design and performance of an OHSP:

1. Previously unknown viruses may present without warning. In fact, in the case of the Ebola-Reston virus, the source of the multiple introductions beyond the single export facility in the Philippines is still not known.
2. Good veterinary staff can note atypical clinical syndromes or death patterns but they are not able to make accurate diagnoses without pathologic and microbiologic support.
3. Some viruses have never been tested by introduction into a closed primate facility, so the epidemiologic patterns may differ from those in other situations. Viruses may change in their epidemiologic patterns or even their genetic capability during the course of transmissions within a primate facility. Parenteral transmission of the Ebola-Reston virus was important initially, but indications of aerosol spread were evident later in the course of the epidemic.
4. Transmission of multiple viruses may occur in the unusual conditions prevailing in the stressful transport, pooling, and housing of nonhuman primates. Co-circulation of simian hemorrhagic fever and the Ebola-Reston virus occurred, and their interrelationship was not understood; but the Ebola-Reston virus alone clearly was capable of causing lethal disease in macaques.
5. Necropsies played an important role in the outbreak. A complete necropsy with tissue blocks permits diagnosis and provides material for virus diagnosis as well

Pox Viruses

Disease Profile in Nonhuman Primates

Five pox viruses have been described as affecting nonhuman primates (Brack 1987; Muchmore 1987; Whitney 1976). Most noteworthy of them is monkeypox because of its clinical and immunologic similarities to smallpox and vaccinia, the other members of the genus *Orthopoxvirus* that affect humans. Outbreaks of human monkeypox occur often in Central and West Africa (Hutin and others 2001) and originally led to some confusion with smallpox virus activity. Monkeypox can experimentally infect a wide variety of Old World and New World nonhuman-primate species (Downie 1974; Heberling and Kalter 1971). Although antibodies may be found in wild populations of guenons, arboreal squirrels are thought to be the natural mammalian reservoirs in endemic regions of Africa (Khodakevich and others 1986), though monkey-to-monkey transmission can occur. Clinically, monkeypox presents as a vesicular exanthema lead-as formalin blocks for future immunohistochemistry. It is also dangerous and should be undertaken only by highly trained persons with proper containment.

6. The level of technical knowledge of staff varied. Communication of risk to illiterate and immigrant workers was difficult because of both language and cultural barriers. In addition, there was a variable response to the idea of risk amelioration by the corporate interests involved. There were adequate numbers of ABSL-4 trained technicians and veterinary staff for many of the procedures once the consulting US Department of Defense laboratory was involved, but this would no longer be true, because of the lessening of ABSL-4 operations and would not be true at most sites around the world.

7. The level of veterinary clinical care was very high, but in spite of this, some practices, particularly recognition of the need for needle hygiene, increased the risks to personnel.

8. State and federal statutes had important effects on the responses that were possible and those implemented. The nonhuman primates were private property, so cooperation had to be requested and could not be mandated. Transport of animal samples a few miles across state lines required multiple time-consuming licenses and permissions. Disposal of animal carcasses was subject to regulations drafted for general environmental protection, and variances had to be sought for environmentally unimportant activities that greatly enhanced biologic safety.

9. An inadequate international technical base with respect to the filoviruses had a severe adverse effect on the response to the emergency.

10. The relatively modest response from the press would probably not be the case today, given the presence of "real-time" cable news.

11. Efficient aerosol spread or severe human disease would have resulted in a different public health scenario.

ing to classic pock-like lesions of the skin and internal organs, sometimes with respiratory involvement.

Yaba virus and tanapox virus (benign epidermal monkeypox or Or-Te-Ca virus) are both members of the *Yatapoxvirus* genus and can cause multiple papules or masses (pseudotumors) on the skin, which occasionally involve oral, pulmonary, and other internal tissues of nonhuman primates, depending on the virus and host species. Evidence of infection has been found in macaques, baboons, guenons, mangabeys, patas monkeys, and chimpanzees, but not in New World monkey species (Brack 1987; Downie 1974). Lesions spontaneously regress by necrosis and ulceration in 3-6 weeks.

Molluscum contagiosum, a skin infection caused by the only member of the *Molluscipoxvirus* genus, causes umbilicated papules on the skin (genital area, face, and upper body) that can persist for up to 2 years. It is a relatively common virus thought to be specific to humans (especially immunosuppressed humans), but cases have been documented in chimpanzees (Douglas and others 1967).

Mode of Transmission

Direct contact and indirect contact (possibly via invertebrate vectors) are suspected but conclusive data are lacking. Transmission of yaba and tanapox from monkeys to humans has been documented on several occasions (including via contaminated needles), and outbreaks among human settlements have occurred in Central Africa with lesions resembling those in animals (Jezek and others 1985). In contrast with yaba and tanapox, person-to-person spread is a feature of monkeypox, although less so than that of smallpox (Hutin and others 2001). To date, nonhuman primates have not been directly implicated in human cases of monkeypox (Hutin and others 2001). Molluscum contagiosum is spread by close direct contact between humans and potentially through close contact with an infected chimpanzee.

Incubation Period and Clinical Signs

The incubation period for monkeypox in humans is thought to be about 3 weeks. Fever precedes pock-like lesions similar to those described above for nonhuman primates. Lesions arise approximately 1 week after human exposure to the *Yatapoxvirus* agents, and published case reports indicate that up to 6 months may be required for molluscum contagiosum to manifest in the patient.

Diagnosis and Prevention

Serology, virus isolation, and PCR assays have been developed for some agents in this group. Monkeypox cases in endemic regions of Africa may have been prevented in previous decades when smallpox (vaccinia) immunization campaigns were active, and none of these should occur in contemporary research settings with nonhuman primates, except perhaps of recently imported ones.

Yellow Fever

Disease Profile in Nonhuman Primates

Yellow fever virus (flavivirus) occurs in sylvatic maintenance cycles involving African Old World primates (including baboons, guenons, mangabeys, chimpanzees, and patas monkeys) and New World primates (including howler monkeys, spider monkeys, capuchins, and squirrel monkeys) (Mansfield and King 1998; Solomon and Mallewa 2001). Transmission of yellow fever virus is via the bite of infective mosquitoes, especially *Aedes aegypti* in urban maintenance cycles and human epidemics, *Ae. africanus* and *Ae. simpsoni* in sylvatic African Old World primate maintenance cycles, and *Haemagogus* spp. in sylvatic New World primate maintenance cycles.

Clinical signs of infection are mild or absent in African species, but New World primates tend to show high mortality, presumably because of the virus's relatively recent introduction into Central and South America in the 16th and 17th centuries.

Mode of Transmission

Human cases acquired from nonhuman-primate research work in laboratory settings have occurred (Richardson 1987), and accidental exposures with needle sticks or other sharp instruments are a continuing possibility, especially during quarantine of animals after receipt from endemic areas.

Incubation Period and Clinical Signs

After an incubation period of 3-6 days, the human disease can present with variable severity, including mild, inapparent infections in endemic areas. However, attacks of rapid onset—including fever, chills, headache, backache, nausea, vomiting, and jaundice—are common. Symptoms can progress to epistaxis and hemoptysis, melena, liver and renal failure, and in some cases, death.

Diagnosis and Prevention

Diagnosis is through serology, virus isolation, PCR, or histopathology. Yellow fever is prevented by maintaining effective quarantine programs, especially after receipt of animals from endemic areas, eliminating or destroying potential larval habitats of *Aedes* mosquito vectors, and protecting against mosquito bites. A human vaccine is available for research workers and the public.

Dengue

Disease Profile in Nonhuman Primates

Dengue fever viruses (flaviviruses) circulate in tropical latitudes among different species of African and Asian Old World primates, including macaques (Mansfield and King 1998; Solomon and Mallewa 2001). The dengue virus is introduced via the bite of infective mosquitoes, especially *Aedes aegypti and Ae. albopictus*. Both nonhuman primates and humans are competent reservoir hosts, and urban transmission cycles occur most regularly between people without dependence on nonhuman primates.

Mode of Transmission

Human cases acquired from nonhuman-primate research work in laboratory settings are not known, but accidental exposures through needles or other medical sharps are possible, especially during periods of quarantine after receipt of animals from endemic areas.

Incubation Period and Clinical Signs

This incubation period in humans is typically 3-7 days until the onset of clinical signs, which generally include a febrile illness accompanied by various signs, such as rash. Severity varies from none to hemorrhagic episodes—and occasional death in the case of dengue hemorrhagic fever-shock syndrome.

Diagnosis and Prevention

Diagnosis is through serologic testing and virus isolation assays. Prevention is by maintaining effective quarantine programs, especially after receipt of animals from endemic areas, eliminating or destroying potential larval habitats of *Aedes* mosquito vectors, and protecting against mosquito bites. No effective vaccine is available.

Lymphocytic Choriomeningitis Virus

Disease Profile in Nonhuman Primates

Marmosets and tamarins in zoological settings have succumbed to epizootics of lymphocytic choriomeningitis virus (LCMV), the etiologic agent of callitrichid hepatitis, after their feeding on infected mice (Montali and others 1989; Stephensen and others 1991). Dyspnea, anorexia, lethargy, jaundice, and high mortality have been reported. Rodents are the natural reservoir species (mice) and long-term carriers (hamsters) of LCMV and are responsible for maintenance cycles of the agent with chronic viremia and viruria. Transmission of the agent between callitrichid primates has not been demonstrated.

Mode of Transmission

The mode of transmission from infected animals to humans is unknown but presumably would result from oral and respiratory exposure to the virus through urine, saliva, or feces of infected animals, as in cases of human exposure to infected rodents. Seroconversion to LCMV has occurred in two zoo veterinarians after bite wounds from and necropsy examinations of infected animals during outbreaks (Montali and others 1989)

Incubation Period and Clinical Signs

The incubation period of LCMV in humans is around 8-13 days. A biphasic febrile illness has most often been reported in association with LCMV infection from rodents. Initial symptoms most often include fever, malaise, anorexia, muscle aches, headache, nausea, and vomiting. The second phase of LCMV includes symptoms of meningitis or encephalitis such as fever, headache, stiff neck, confusion, or motor abnormalities. LCMV is only rarely fatal, though temporary or permanent neurological damage is possible. Recent reports regarding the role of LCMV as a fetal teratogen raise concerns about exposure of women of childbearing age to this agent (Barton and Mets 2001).

Diagnosis and Prevention

Serology, virus isolation assays, and PCR are used for diagnosis. Screening of rodent colonies that are used as food for callitrichids and prevention of access by wild rodents should be sufficient to prevent transmission. Research or veterinary biologics for callitrichid species should be verified as free of LCMV before use. Barrier methods of protection and safe-handling procedures in work with all nonhuman primates should also minimize the likelihood of occupationally acquired infections.

Hepatitis A

Disease Profile in Nonhuman Primates

Three genotypes of simian hepatitis A virus have been found naturally in Old World primate species, including macaques and guenons (Robertson 2001). Serologic evidence of infection with simian hepatitis A has been seen in macaques, baboons, guenons, cebid and callitrichid monkeys, and other species (Lankas and Jensen 1987; Mansfield and King 1998). Molecular characterization of simian isolates shows limited nucleotide-sequence identity with human or other known isolates (Oberste and others 2002; Poyry and others 1999), and there is no evidence of human disease in association with these simian agents.

Many species of nonhuman primates are susceptible to hepatitis A viruses of human origin, and tamarins, owl monkeys, and chimpanzees have been valuable animal models of the disease (Purcell and Emerson 2001). Outbreaks of human hepatitis A in captive colonies of nonhuman primates have been attributed to exposure to human strains after capture, but only chimpanzees have been implicated in retransmission of the human hepatitis A virus to humans (Dienstag and others 1976; Robertson 2001). Overt clinical disease in nonhuman primates after infection has been rare and nonspecific; reported signs include anorexia and diarrhea in chimpanzees, although increases in liver enzyme levels to 2-10 times above normal have been noted. Hepatitis A virus may persist in chimpanzee serum for up to 91 days—much longer than previously suspected (Bower and others 2000; Cohen and others 1989).

Mode of Transmission

Transmission of hepatitis A to human beings is usually by a fecal-oral route, although transmission via parenteral inoculation is possible.

Incubation Period and Clinical Signs

The incubation period of human hepatitis A (potentially retransmitted through nonhuman primates) averages about 1 month. Illness is generally self-limiting. Signs include fever, malaise, anorexia, nausea, abdominal discomfort, and jaundice. Many cases are asymptomatic. There is no evidence of human disease associated with simian hepatitis A.

Diagnosis and Prevention

Serology is routinely used for diagnosis. Virus isolation with genomic analysis would be required for species-associated genotype investigations. Good hygiene, barrier methods of protection, and safe-handling procedures help to prevent transmission from animals infected experimentally or inadvertently with human-origin isolates. A vaccine is available, and it may be advisable to vaccinate persons who work with hepatitis A in nonhuman-primate research settings and persons who are occupationally exposed to chimpanzees. However, since hepatitis A strains from humans and nonhuman primates differ genetically and there is no evidence one way or the other that the current human vaccine would protect humans from simian strains of this virus, it is unclear as to whether vaccinations would be advantageous.

Hepatitis B

Disease Profile in Nonhuman Primates

Hepatitis B virus (HBV) infection can occur in the great apes (chimpanzees, gorillas, orangutans, and gibbons) with moderate frequency; this is due in some cases to previous exposure of the animals to human serum and in others to naturally occurring genotypes. Species-specific strains of HBV virus have been found in chimpanzees, gibbons, orangutans, and woolly monkeys; isolates from the latter species are phylogenetically most divergent from the rest. (Heckel and others 2001; Robertson 2001; Robertson and Margolis 2002; Takahashi and others 2000). Chronic-carrier status has been established for experimentally infected chimpanzees, and the clinical signs have been limited to mild anorexia, jaundice, and elevated liver enzyme levels. No transmission of these primate-origin agents to humans has been documented, but antigenic studies of the chimpanzee HBV suggest that it could infect humans (Robertson 2001). Laboratory accidents (needle sticks, mucosal exposures, possibly animal bite wounds) could pose HBV hazards for persons involved with animal models of the agent.

Mode of Transmission

No cases of transmission of HBV of nonhuman primate origin to humans has been documented, though these agents could presumably be spread by percutaneous inoculation, permucosal exposure to infective material, or vertically via mother to child in the case of human-to-human HBV.

Incubation Period and Clinical Signs

The incubation period of human HBV averages about 2-3 months. Most infections in humans go unrecognized, but various nonspecific sign —such as anorexia, abdominal discomfort, nausea, and vomiting—may be seen and often progress to jaundice. Chronic cases can persist as active hepatitis, with or without cirrhosis, and progress to hepatocellular carcinoma.

Diagnosis and Prevention

Serum antigen and antigen-assay systems are available and are directed at different markers of HBV infection that can help to indicate the type and duration of infection. Good hygiene, barrier methods of protection, and safe-handling procedures help prevent occupational exposures in laboratory settings, and the vaccines now available for the public are a requirement of the Occupational Health and Safety Administration (OSHA) for the protection of persons involved in research work on this agent and should be considered as well for all persons who work with chimpanzees or their blood, organs, or other tissues.

BACTERIAL DISEASES

Tuberculosis

Disease Profile in Nonhuman Primates

Mycobacterium tuberculosis and *M. bovis* can be acquired from humans and then passed between nonhuman primates. These agents can infect any species of nonhuman primate (CDC 1993; Fourie and Odendaal 1983; Kalter and others 1978; Kaufmann and others 1975; Mayhall and others 1981; Muchmore 1987; Renquist 1987; Richardson 1987; Rollin and others 1999; Sapolsky and Else 1987; Whitney 1976; Zumpe and others 1980), and secondary spread back to humans has been documented on several occasions. Old World nonhuman-primate species are more susceptible to tuberculosis infection than New World nonhuman-primate species (Aiello 1998).

There is no apparent difference between the two agents in the presentation of disease in nonhuman primates. Signs are often nonspecific and vary with the location and severity of infection, from asymptomatic cases to overwhelming infection with sudden death. They include pulmonary disease that may or may not be outwardly apparent, anorexia, and chronic weight loss. Chronic draining tracts to the skin and vertebral bone and spinal-cord involvement have been reported.

Mycobacterium avium-intracellulare and other atypical agents in this group can occur naturally in nonhuman primates, especially if they are immunocompromised; however, these are generally considered ubiquitous in nature, and nonhuman primates are not considered a reservoir source for human infections (Gibson 1998).

Mode of Transmission

Aerosols, fomites, and the fecal-oral route have all been implicated in recorded outbreaks among nonhuman primates.

Incubation Period and Clinical Signs

The incubation period for tuberculosis in humans is about 2-10 weeks from exposure to development of primary lesions or skin-test positivity. Pulmonary, meningeal, and other body systems can be involved. Signs can include chronic cough, fatigue, fever, weight loss, and hemoptysis during advanced stages of progressive pulmonary disease. Lifelong latent (nonprogressive) infections in calcified pulmonary nodules are a known feature.

Diagnosis and Prevention

Intradermal skin tests with mammalian old tuberculin, accompanied by pathologic examination as necessary, are the mainstay of tuberculosis surveillance in nonhuman-primate colonies. An enzyme immunoassay for detection of tuberculosis infection in various species of nonhuman primates via measurement of gamma interferon release from stimulated lymphocytes is also available (Desem and Jones 1998). Intradermal skin tests using PPD and pulmonary radiography, coupled with acid-fast staining and culture of sputum samples are used for humans. However, false-negative skin test results in nonhuman primates are common because of concurrent disease (such as measles virus infection), early-stage disease, or other poorly understood factors and justify the requirement for repeated testing during and after quarantine.

Test-and-removal and depopulation strategies have been used to control the disease in colonies, and must be combined with rigorous quarantine programs and continuing tuberculosis surveillance of nonhuman primates and persons who work with them to prevent colony infection. Good sanitation programs for facilities and equipment can limit spread in a colony. Some long-term multidrug protocols have been described and have been successful in individual nonhuman-primate cases. Good personal hygiene, respiratory protection for persons working with nonhuman primates not known to be free of active infection, safe-handling procedures, and rigorous quarantine and surveillance programs can minimize spread in a colony. Prevention of contact of actively infected persons with nonhuman primates through appropriate occupational health programs is also paramount.

Pseudotuberculosis

Disease Profile in Nonhuman Primates

Pseudotuberculosis, caused either by *Yersinia enterocolitica* or *Y. pseudotuberculosis,* has been reported in macaques, baboons, squirrel monkeys, guenons, and other nonhuman primates (Bielli and others 1999; Brack and Hosefelder 1992; Gibson 1998; Plesker and Claros 1992; Vore and others 2001). Birds and rodents are thought to be the primary reservoir, but nonhuman primates may be carriers and could present a risk of infection to humans.

Diarrhea, dehydration, anorexia, and weight loss are noted, sometimes hemorrhaging and occasionally splenomegaly and lymphadenopathy are seen. A peracute course ending in death has also been observed.

Mode of Transmission

Transmission is by the fecal-oral route in both humans and nonhuman primates.

Incubation Period and Clinical Signs

Incubation is about 3-7 days and is followed by acute febrile diarrhea, which may be hemorrhagic; enterocolitis; and mesenteric lymphadenitis that mimics appendicitis in infected persons may also be seen.

Diagnosis and Prevention

Diagnosis is through stool culture with biochemical evaluation of the isolates. Good hygiene, barrier methods of protection, and safe-handling procedures are preventive. Antimicrobial treatment of infected animals has been discouraging.

Shigellosis

Disease Profile in Nonhuman Primates

Shigella appears to be acquired by nonhuman primates in captivity from contact with infected humans, and various species of *Shigella,* most commonly *S. flexneri,* have been found in association with a variety of nonhuman-primate species, including macaques, baboons, squirrel monkeys, marmosets, tamarins, and apes (Gibson 1998; Juan-Salles and others 1999; Miller and others 1990; Muchmore 1987; Wolfensohn 1998). Transmission of the agent back to humans has been seen on multiple occasions. The presence of asymptomatic, chronic carriers and reinfections can maintain high rates of endemic infections in research colonies; stress promotes episodes of overt disease. Clinical signs can include diarrhea containing mucus, blood, and sometimes mucosal fragments; dehydration; and weight loss. Gingivitis, abortion, and air sacculitis are also reported.

Mode of Transmission

Transmission is by the fecal-oral route in humans and nonhuman primates.

Incubation Period and Clinical Signs

In humans, the incubation period averages 1-4 days. The disease varies from mild infections to dysentery or watery diarrhea, fever, nausea, and tenesmus. It can cause reactive arthropathy (Reiter's syndrome), especially in persons with an HLA-B27 genetic background (Hughes and Keat 1994).

Diagnosis and Prevention

Diagnosis is through stool culture and biochemical typing of isolates. Good hygiene, barrier methods of protection, and safe-handling procedures are preventive. Infected animals should be treated if possible. Culling of chronic carriers that do not respond to therapy has been used in some primate colonies.

Salmonellosis

Disease Profile in Nonhuman Primates

As in the case of *Shigella*, the *Salmonella* pathogen is probably acquired by nonhuman primates in captivity from exposure to infected humans, and transmission of the agent back to humans is not unlikely. Various species and bioserotypes of *Salmonella* have been isolated from nonhuman primates such as macaques, guenons, tamarins, owl monkeys, and chimpanzees (Gibson 1998; Muchmore 1987; Takasaka and others 1988). However, it is rarely reported in established colonies. Watery diarrhea, sometimes containing blood or mucus; fever; and such extraintestinal infections as neonatal septicemia, abortion, osteomyelitis, and pyelonephritis are described in association with *Salmonella* infections in these species.

Mode of Transmission

Transmission is by the fecal-oral route in both humans and nonhuman primates.

Incubation Period and Clinical Signs

The incubation period is 1-3 days in humans. The disease typically presents as acute enterocolitis with fever, abdominal pain, headache, nausea, and vomiting. Localization in various tissues of the body may occur and induce abscesses and septicemia.

Diagnosis and Prevention

Diagnosis is through stool culture and biochemical typing of isolates. Good hygiene, barrier methods of protection, and safe-handling procedures are preventive. Culling of infected animals has been used in some primate colonies.

Campylobacteriosis

Disease Profile in Nonhuman Primates

It is likely that many nonhuman-primate cases of *Campylobacter jejuni* and *C. coli* are acquired in captivity; *Campylobacter* has been commonly found in association with macaques, baboons, squirrel monkeys, cebid monkeys, chimpanzees, tamarins, and other nonhuman primates (Gibson 1998; Hubbard and others 1991; Johnson and others 2001; Renquist 1987; Taylor and others 1989; Tribe and Frank 1980). Retransmission of *Campylobacter* back to humans is not unlikely. Diarrheal disease of varied severity and duration may be seen, but many infections are asymptomatic.

Mode of Transmission

Transmission is by the fecal-oral route in humans and nonhuman primates.

Incubation Period and Clinical Signs

The incubation period is typically 2-5 days in humans. Diarrhea, abdominal pain, fever, nausea; and vomiting have been seen; stools sometimes contain blood or mucus.

Diagnosis and Prevention

Diagnosis is through stool culture with morphologic evaluation (the agent is spiral shaped) and biochemical evaluation of the isolates. Good hygiene, barrier methods of protection, and safe-handling procedures are preventive. Infected animals should be treated if possible, although the efficacy of antibiotic therapy is infrequently reported and some animals continue to shed the organism despite treatment (Aiello 1998).

Melioidosis

Disease Profile in Nonhuman Primates

Burkholderia (Pseudomonas) pseudomallei has been found in macaques, chimpanzees, and other nonhuman primates with nonspecific clinical signs including suppurative lesions involving pulmonary, cerebrospinal, and subcutaneous tissues (Fritz and others 1986; Gibson 1998; Trakulsomboon and others 1994;). The zoonotic potential of this agent has not been fully established. *B. pseudomallei* is considered a soil and water saprophyte in tropical countries and most cases have arisen among animals recently imported from endemic areas.

Mode of Transmission

Laboratory-acquired infections have occurred, and infection is thought to occur through ingestion or inhalation or via breaks in the skin.

Incubation Period and Clinical Signs

Long latent periods are characteristic. Human cases vary from asymptomatic infections to pneumonia and septicemia.

Diagnosis and Prevention

Diagnosis is by culture of the agent from affected tissues. Good hygiene, barrier methods of protection, and safe-handling procedures are warranted to minimize likelihood of exposure from infected animals. Antibiotic therapy has been discouraging in nonhuman primates because of bacterial resistance.

Leprosy

Disease Profile in Nonhuman Primates

Mycobacterium leprae has been recorded in macaques, mangabeys, and chimpanzees (Hubbard and others 1991; Meyers and others 1991; Valverde and others 1998) with single or multiple nodular lesions of the face and ears sometimes extending to other areas. Infections are believed to be acquired naturally in endemic areas. Weight loss and anemia develop late in disease.

Mode of Transmission

The exact mode of transmission is not well established, though it is suspected that most human infections are acquired via inhalation.

Incubation Period and Clinical Signs

Months to years follow exposure before onset of clinical signs in humans. Various forms of disease are reported, from tuberculoid to lepromatous with chronic infiltrative skin and nerve involvement. Zoonotic cases originating in infected nonhuman primates have not been recorded, but the potential exists.

Diagnosis and Prevention

Diagnosis is through histopathology with acid-fast staining of lesions, which is especially convincing when there is nerve involvement. Good hygiene, barrier methods of protection, and safe-handling procedures are warranted. Early recognition and isolation of cases, possibly including long-term multidrug therapy, are also beneficial.

Bite-Wound Infections

Apart from the conditions listed above, there is a potential for infectious agents to complicate bite injuries sustained from nonhuman primates, as is the case for any animal in general. The severity of B virus from macaques can overshadow these concerns during care and follow-up of bite-wound cases, but the likelihood of bacterial wound contamination after bites by nonhuman primates also warrants clinical attention with early initiation of antimicrobial therapy. The list of bacteria of interest may include *Neisseria* spp., *Streptococcus* spp., *Staphylococcus* spp., *Haemophilus parainfluenzae, Eikenella corrodens, Moraxella* spp., *Bacteroides* spp., *Fusobacterium* spp., *Clostridium tetani*, and *Pasteurella multocida*; some fungal agents can also contaminate bite wounds (Goldstein and others 1995; Janda and others 1990; Rayan and others 1987; Tribe and Noren 1983). The tetanus-immunization status of injured persons should be ascertained immediately and boosted as necessary. Reviews of animal-bite wound cleansing and medical management in general should be consulted (Smith and others 2000). Both aerobic and anaerobic cultures should be taken and empirical therapy with antimicrobial agents initiated as appropriate.

PROTOZOAN PARASITES

Amoebiasis

Disease Profile in Nonhuman Primates

Several species of *Entamoeba* occur in nonhuman primates, but the only known pathogenic species is *Entamoeba histolytica*. *Entamoeba histolytica* is reported in macaques, baboons, squirrel monkeys, guenons, mangabeys, and chimpanzees (Ghandour and others 1995; Hubbard and others 1991; Levine 1970; Munene and others 1998; Muriuki and others 1998; Sargeaunt and others 1982; Smith and Meerovitch 1985). The presence of diarrhea (none to severe) and weight loss (dysentery) depends on the strain of organism. Diarrhea may be hemorrhagic or catarrhal.

Mode of Transmission

Transmission is by the fecal-oral route in humans and nonhuman primates. Human cases associated with animal contact are rare.

Incubation Period and Clinical Signs

The incubation period is variable but typically ranges from 2 to 4 weeks. Many infections are asymptomatic, but acute or fulminating dysentery with fever, chills, and hemorrhagic or catarrhal diarrhea may occur.

Diagnosis and Prevention

Fecal wet-smear examination to demonstrate trophozoites and serologic tests for invasive forms of the disease are available. Antigen capture assays are also available. No successful treatment for eradication of *E.*

histolytica in nonhuman-primate colonies has been reported, therefore colony screening for *E. histolytica* may be warranted (Weber and others 1999). Good hygiene, barrier methods of protection, and safe-handling procedures are indicated.

Balantidiasis

Disease Profile in Nonhuman Primates

Balantidium coli is reported in macaques, baboons, squirrel monkeys, guenons, and chimpanzees (Ghandour and others 1995; Hubbard and others 1991; Knezevich 1998; Levine 1970; Munene and others 1998; Muriuki and others 1998; Nakauchi 1999). Disease manifestations vary from none to watery diarrhea and ulcerative enterocolitis, weight loss, and rectal prolapse.

Mode of Transmission

Transmission is by the fecal-oral route in humans and nonhuman primates. Human cases associated with animal contact are rare.

Incubation Period and Clinical Signs

The incubation period is undefined but likely to be only a few days. Diarrhea, tenesmus, nausea, and vomiting are described for human infections.

Diagnosis and Prevention

Fecal examination is used to demonstrate trophozoites. Good hygiene, barrier methods of protection, and safe-handling procedures are indicated.

Cryptosporidiosis

Disease Profile in Nonhuman Primates

Cryptosporidium parvum is reported in macaques, baboons, squirrel monkeys, guenons, chimpanzees, and marmosets (Miller and others 1990; Muriuki and others 1998; Toft and Eberhard 1998). Disease manifestations vary from none to gastroenteritis and intractable diarrhea, dehydration, and weight loss. The disease is fairly common in macaques that are experimentally infected with SIV.

Mode of Transmission

Transmission is by the fecal-oral route in humans and nonhuman primates. Human cases associated with animal contact are rare.

Incubation Period and Clinical Signs

The incubation period is likely to be 1-12 days. Abdominal cramping, watery diarrhea, nausea, and vomiting are described for human infection.

Diagnosis and Prevention

Fecal examination is used to demonstrate oocysts; antigen capture assays are also available. Good hygiene, barrier methods of protection, and safe-handling procedures are indicated.

Giardiasis

Disease Profile in Nonhuman Primates

Giardia intestinalis is reported in macaques, baboons, squirrel monkeys, chimpanzees, marmosets, and other nonhuman primates (Ghandour and others 1995; Hamlen and Lawrence 1994; Levine 1970; Toft and Eberhard 1998). Disease manifestations vary from asymptomatic to diarrhea and vomiting, depending on a range of poorly understood factors.

Mode of Transmission

Transmission is by the fecal-oral route in humans and nonhuman primates. Human cases directly associated with animal contact are rarely documented.

Incubation Period and Clinical Signs

The incubation period can be prolonged to about 3 weeks but is typically 7-10 days. Many infections are asymptomatic, but chronic diarrhea, steatorrhea, abdominal cramps, bloating, and fatigue are described.

Diagnosis and Prevention

Fecal examinations are used to demonstrate cysts or trophozoites, and antibody-based antigen-detection systems for stool specimens are available. Good hygiene, barrier methods of protection, and safe-handling procedures are indicated.

Malaria

Disease Profile in Nonhuman Primates

Plasmodium cynomolgi, P. knowlesi, P. inui, P. simium, and other species have been described in macaques, baboons, squirrel monkeys, mangabeys, and other nonhuman primates (Bennett and McWilson 1965; Collins and others 1973; Most 1973; Muchmore 1987; Ollomo and others 1997; Toft and Eberhard 1998; Trakulsomboon and others 1994). Clinical signs are typically absent in the natural host species, but slight anemia in conjunction with low-grade parasitemia may occur. Stress, concurrent disease, splenectomy, or immunosuppression can precipitate episodes of overt disease in infected animals.

Mode of Transmission

Human cases acquired directly from nonhuman primates are rare; accidental exposures through penetrating injuries from needles and other medical sharps are possible as is transmission through a mosquito bite.

Incubation Period and Clinical Signs

The incubation period in humans is typically 1-4 weeks after the bite of an infective female mosquito of the genus *Anopheles*, which may be followed by fever, chills, sweating, headache, and nausea.

Diagnosis and Prevention

Blood-smear evaluations, PCR, and serologic tests for past infections are available. Safe blood-handling practices and effective mosquito-control programs are indicated in work with infected animals.

Trypanosomiasis

Disease Profile in Nonhuman Primates

Natural infections with *Trypanosoma cruzi* and related *Trypanosoma* organisms have been described in squirrel monkeys, marmosets, tamarins, and other New World species, in which infections are apparently lifelong. Infections have also been reported in macaques, baboons, and great apes, mostly in captive colonies in endemic areas where infection may have been acquired locally (Levine 1970; Ndao and others 2000; Toft and Eberhard 1998). Various nonspecific clinical signs are associated with infection; myocarditis is the most common.

Mode of Transmission

Human cases acquired from nonhuman-primate research work are not known, but accidental exposures through mucous membranes, non-intact skin, or penetrating injuries from needles and other medical sharps are possible. *T. cruzi* can be propagated among monkeys in closed-colony settings through trauma, blood-to-blood exposure, saliva, sexual activity, and transplacentally (Ndao and others 2000).

Incubation Period and Clinical Signs

The incubation period in humans is typically 1-2 weeks. Fever, malaise, lymphadenopathy, hepatosplenomegaly, myocardial damage, and a constellation of other signs are known features of this disease (Chagas disease) in humans.

Diagnosis and Prevention

Blood-smear evaluations, blood culture, serologic tests, and PCR assays are available. Safe blood-handling practices and effective control programs for involved vector species are indicated in work with infected animals.

METAZOAN PARASITES

Hymenolepiasis

Disease Profile in Nonhuman Primates

The cestode *Hymenolepis nana* has been reported in many species of nonhuman primates, including macaques, baboons, squirrel monkeys, and chimpanzees (Ghandour and others 1995; Muchmore 1987; Toft and Eberhard 1998). Most infections are asymptomatic, but catarrhal enteritis, diarrhea, and abdominal signs have been reported.

Mode of Transmission

Both a direct and an indirect life cycle involving insect vectors are possible. Fecal-oral exposures are plausible, in that eggs are infective when passed. Human cases acquired directly from nonhuman-primate research work have not been reported, but accidental exposures are plausible.

Incubation Period and Clinical Signs

The incubation period is variable—about 2 weeks to maturation of worms in the host. Minor infections are asymptomatic, but enteritis with or without diarrhea, abdominal pain, weight loss, and other symptoms have been reported.

Diagnosis and Prevention

Feces should be examined for eggs. Good hygiene, barrier methods of protection, and safe-handling procedures are indicated. Infected animals should be treated with anthelmintics, and intermediate vector insects should be eliminated.

Oesophagostomiasis

Disease Profile in Nonhuman Primates

The nematodes *Oesophagostomum apiostomum* and *O. bifurcum* have widespread distribution and are found occasionally in macaques, baboons, guenons, mangabeys, and chimpanzees (Abbott and Majeed 1984; Muchmore 1987; Munene and others 1998; Perolat and others 1992; Toft and Eberhard 1998;). Infected animals may be asymptomatic or can show a failure to thrive, weight loss, and diarrhea. Nodules containing viable or dead worms are seen on the serosal surface of bowel and other organs, and colonic ulcers may occur.

Mode of Transmission

Transmission is by the fecal-oral route in humans and nonhuman primates.

Incubation Period and Clinical Signs

The incubation period is poorly defined but is probably weeks to months. Abdominal pain and tenderness, appendicitis, and nodular inflammation of the intestinal wall have been described.

Diagnosis and Prevention

Fecal examination, serology, and ultrasonography for identification of nodules are warranted. Adult worms are required for definitive identification to avoid potential confusion with other nematodes. Good hygiene, barrier methods of protection, and safe-handling procedures are indicated, as is anthelmintic treatment of any infected animals.

Oxyuriasis

Disease Profile in Nonhuman Primates

Oxyuriasis is a group of threadworms that includes the pinworm nematode *Enterobius vermicularis.* This parasite occurs in Old World monkeys and apes, including baboons, guenons, macaques, and chimpanzees (Brack 1987; Ghandour and others 1995; Hubbard and others 1991; Muchmore 1987; Munene and other 1998; Toft and Eberhard 1998). Clinical signs in these species are generally absent. There has been some mention of anal pruritus and irritation, and severe infections have occasionally led to progressive enterocolitis, peritonitis, and death in chimpanzees. Nonhuman primates can acquire this agent from contaminated soil in endemic regions; it can then be passed between species in either direction.

Mode of Transmission

Transmission in humans and nonhuman primates is by the fecal-oral route. Eggs are deposited around the anus and become infective within a few hours, sometimes appearing in the feces. Successive reinfections can occur by transfer of eggs to the host's mouth (accidentally or via coprophagy in some species). Dustborne infections may occur in heavily contaminated environments.

Incubation Period and Clinical Signs

The life cycle of oxyuriasis is about 2-6 weeks. It takes several months for a person to develop a high parasite burden leading to symptomatic disease. People may report perianal itching, disturbed sleep, irritability, and occasional secondary skin infections of the perianal region from self-caused trauma.

Diagnosis and Prevention

A perianal tape test for eggs is standard technique. Fecal examination or sigmoidoscopy of the lower colon can also be done. Good hygiene, barrier methods of protection, and safe-handling procedures are indicated, and infected animals should be treated appropriately with anthelmintics.

Strongyloidiasis

Disease Profile in Nonhuman Primates

The nematode *Strongyloides fullerborni* is reported in macaques, baboons, guenons, mangabeys, chimpanzees, and other nonhuman primates (Abbott and Majeed 1984; Battles and others 1988; Hubbard and others 1991; Knezevich 1998; Muchmore 1987; Munene and others 1998; Muriuki and others 1998; Toft and Eberhard 1998;), and *Strongyloides stercorales* has been documented in apes and other nonhuman primates (Penner 1981). Diarrhea (sometimes hemorrhagic or mucoid) is most commonly reported and is accompanied by weight loss, anorexia, vomiting, coughing, pulmonary hemorrhage, and other signs including death in the case of severe infections.

Mode of Transmission

Transmission in humans and nonhuman primates is by the fecal-oral route, as well as by direct skin penetration for larval stages. Free-living larval forms of this parasite are described.

Incubation Period and Clinical Signs

The incubation period is uncertain, but probably 2-4 weeks. Abdominal pain, nausea, diarrhea, and anemia are reported symptoms of *Strongyloides* infection in humans.

Diagnosis and Prevention

Fecal examination for eggs and larvae is standard technique. Good hygiene, barrier methods of protection, and safe-handling procedures are indicated, especially in light of the potential for free-living forms in facilities where nonhuman primates are housed. Infected animals should be treated appropriately with anthelmintics.

Trichuriasis

Disease Profile in Nonhuman Primates

The whipworm nematode *Trichuris trichuria* and its close relatives have a worldwide distribution in New World and Old World monkeys and the great apes (Brack 1987; Ghandour and others 1995; Hubbard and others 1991; Knezevich 1998; Toft and Eberhard 1998). Minor infections are typically asymptomatic; severe infections can cause anorexia, diarrhea, enteritis, and occasionally death.

Mode of Transmission

Transmission in humans and nonhuman primates is by the fecal-oral route but requires 10-14 days of incubation in warm, moist soil to become infective.

Incubation Period and Clinical Signs

The period to onset of clinical signs is not well established, but symptoms begin before the appearance of eggs in the feces, which takes about 3 months following ingestion.

Diagnosis and Prevention

Fecal examination for eggs and sigmoidoscopy of the lower colon are typically used. Good hygiene, barrier methods of protection, and safe-handling procedures are indicated, and infected animals should be treated appropriately with anthelmintics.

OTHER AGENTS

Because of the diversity of species involved and shared susceptibilities to common pathogens, many other infectious agents occasionally found in nonhuman primates are of potential or documented risk to persons who work with them. A few examples are given here, but other sources should be consulted for situations where comprehensive listings are needed (CDC-NIH 1999; NRC 1996).

There are a number of reports of rabies virus infections of New World monkeys, Old World monkeys, and apes; all nonhuman primates should be considered susceptible (Brown 1997; Richardson and Humphrey 1971; Whitney 1976). Both furious and paralytic forms have been seen in these species but without clinical signs specific to the condition. Human cases of rabies from nonhuman primates are generally rare, although eight deaths due to a newly described rabies virus variant have recently occurred after bite injuries from infected pet marmosets in Brazil (Favoretto and others 2001). Nonhuman primates may be exposed to rabies through bites by infected bats, dogs, or other reservoir species in endemic areas or while they are held before export. Killed vaccines have unknown efficacy when used to protect nonhuman primates, although they have been used on occasion for outdoor-housed animals in rabies-epizootic areas. Barrier methods of protection, safe-handling procedures, and prompt and appropriate follow-up are always warranted in the event of bites or scratches by these species.

Considerable controversy exists regarding the potential that exogenous simian type D retroviruses (SRVs) can be transmitted to humans. SRV occurs naturally in wild and captive macaques, where it is associated with an AIDS-like immunosuppressive disease, retroperitoneal fibromatosis, opportunistic infections, persistent refractory diarrhea, and coma. Problems in addressing this issue have been attributable to the population of persons available for testing and inconclusive seroreactivity in the results. However, one recent study provides strong evidence that SRV should be considered a zoonotic agent based on persistent seropositivity spanning 3 years in one animal handler and seroconversion possibly indicative of a transient infection in another over a 2-year period (Lerche and others 2001). Both handlers had sustained occupational exposure to different species of nonhuman primates during their careers, and both remained healthy despite infection. Use of barrier methods of protection and safe-handling procedures, including prompt disinfecting and flushing of wound sites as warranted because of B virus hazards, should help to minimize transmission in occupational settings.

Recent reports of SV40-specific antigen sequences in some types of non-Hodgkin lymphoma have renewed concerns regarding the zoonotic potential of this agent, inadvertently introduced into the human population via contaminated lots of poliovirus vaccine manufactured from 1955 to 1961 (Malkin 2002; Vilchez and others 2002). There are also epidemiologic links between SV40 infection and human malignant mesothelioma (Carbone and others 2000) and a host of other diseases (Strickler and Goedert 1998).

SV40 is a polyoma virus naturally associated with macaques that can be experimentally transmitted to African Old World species (Brack 1987). Latent infections are established, including in kidney cells used in the original production of poliovirus vaccine. Virus can spread through urinary and nasopharyngeal secretions of infected monkeys and humans (Brack 1987). The extent to which SV40 is biologically important for any human disease remains controversial (Ferber 2002), and there is a vast body of literature on experimental work with this agent in laboratory rodent models. The recommended safety precautions of good hygiene, barrier methods of protection, and safe-handling procedures should help to prevent accidental exposure in nonhuman-primate research settings.

Cases of tularemia (*Francisella tularensis*) have occurred in research colonies of squirrel monkeys (Doyle and others 1988) and other species, without documented sources of infection. Mammalian reservoirs for tularemia in the United States are wild rabbits and rodents, but secondary transmission to nonhuman-primate research workers in this context is conceivable.

Mycoplasma hominis, *Ureaplasma urealyticum*, and other unidentified mycoplasmas are commonly carried by macaques, baboons, and chimpanzees (Schoeb and others 1997), but the zoonotic significance of these agents is unknown.

Naturally acquired cases of *Helicobacter pylori*, *H. heilmannii*, and novel species in this genus have been found in macaques and baboons (Fox and others 2001; Ho and others 1991; Reindel and others 1999) with and without clinically associated disease, adding potential complications to use of infected animals as experimental models for these agents. The zoonotic potential of these agents in the context of biomedical-research use of nonhuman primates remains unresolved. Some success with elimination of *H. pylori* in rhesus monkeys has been documented (Dubois and others 1998). There is also evidence of infection with various serovars of *Leptospira interrogans* in several species of nonhuman primates, including macaques, baboons, squirrel monkeys, guenons, chimpanzees, and tamarins (Fear and others 1968; Gibson 1998; Perolat and others 1992). The agent has been associated with outbreaks of abortion and stillbirth in baboons and abortion and peracute death in squirrel monkeys, but overt disease in other cases of infection has been inapparent. *Leptospira* organisms can be acquired via ingestion, mucous membrane exposure, or skin abrasion; nonhuman primates carrying them could represent a hazard for humans, even if they are not the primary vertebrate reservoir.

There are occasional reports of dermatophytes and ectoparasites (lice and skin mites) with documented or suspected zoonotic significance in various species of nonhuman primates (Baker and others 1971; Gibson 1998; Goldman and Feldman 1949; Gugnani 1971; Ronald and Wagner 1973). Rapid identification and treatment in research settings and the use of barriers covering all exposed skin surfaces should help to reduce the likelihood of occupational exposures.

SUMMARY

Nonhuman primates and their tissues can become infected with micro-organisms communicable to humans in many ways. Exposure of nonhuman primates to zoonotic agents can occur in their native habitats, whether or not they function as ecologically important reservoir hosts. They may also occur through association with humans or human waste material at any stage of their handling and use; through exposure to other infected nonhuman-primate species or other infected vertebrate or invertebrate species during shipping, holding, or use; and through purposeful or inadvertent infection in the course of experimental work. Many of the infectious hazards described in this chapter are considered acutely or chronically pathogenic in nonhuman primates, and well-structured programs of veterinary care should help to identify and eliminate many of these agents from research colonies. Others typically exist for long periods as asymptomatic infections and require special efforts in disease surveillance, often with limited therapeutic options for their complete eradication. Whether these infections can be sustained in research-colony settings depends on the species of nonhuman primate, the biology of the agent, systems of husbandry and veterinary care, and the presence of competent invertebrate vectors. Likewise, the potential for occupational exposure to the agents varies with the collection of species, the type of research use, systems of husbandry and veterinary care, and contact with other vertebrate and invertebrate species at each institution. Other chapters of this book contrast the exposure rates of persons who work with nonhuman primates in a risk-based context and provide a framework for developing and monitoring programs of safety appropriate to these concerns in the modern era.

4

Identifying Noninfectious Hazards

While it is essential to underscore the consideration of infectious hazards, it is also imperative to note that work with nonhuman primates also exposes personnel to various physical and chemical hazards. Few published articles describe noninfectious hazards associated with nonhuman-primate facilities, although potentially they can cause serious injuries and occupational illnesses. It is important for nonhuman-primate workers and managers to be aware of all hazards, not just infectious ones.

Job hazards can be identified in several ways. Epidemiologic investigations have an important role in understanding industry-wide trends and risk factors (Robertson 1998). Most employers are required to maintain a log of workplace injuries and illnesses. Those logs and incident-investigation reports provide useful information to managers when they are evaluating causes of injury and workplace trends. Such evaluation provides opportunities for reducing injuries and exposures. Comprehensive information on identification of hazards in laboratory and research animal settings is available (NRC 1995, 1997), and there are widely accepted principles and guidelines for establishing and implementing effective occupational health and safety programs (International Labour Office 1998; NSC 1996, 1997). These topics are discussed further in Chapters 6 and 7.

The following descriptions of hazards and sources of exposures, injuries, and illnesses (summarized in Tables 4-1 and 4-2) are provided as a tool for individuals whose responsibility is to identify the hazards to personnel who work with nonhuman primates. The identification of the

TABLE 4-1 Summary of Physical Hazards

Hazard Type and Operation	Precautionary Measures
Animal bites and scratches. Animal handling, husbandry	Tranquilize, anesthetize, train, or restrain animals to prevent bite and scratch injuries; use heavy leather or Kevlar® gloves and arm covers; remain at safe working distance from animals; clean all wounds immediately, obtain medical attention
Needle and other wounds from sharp objects. Surgical procedures, phlebotomy, injections	Use safe needle systems when possible; do not recap needles; dispose of contaminated sharp devices in rigid waste container for disposal; clean wound site immediately, obtain medical attention
Slips, trips, and falls. Work in animal housing, cage-wash, other areas	Maintain good housekeeping, remove objects that could cause tripping and liquid or oily slip hazards; use slip-resistant footwear, flooring; maintain all stairs and elevated work platforms in good condition; use carts or dollies to transport animals or heavy objects; plan work so that heavy loads are not carried
Overexertion and repetitive strain injuries. Performing animal, cage, material handling	Do not exceed heavy-lift limits; design equipment and tasks to avoid awkward postures and motions and high repetitions; seek medical attention if early signs of musculoskeletal injury persist
Traumatic crushing and laceration hand injuries. Moving cages, other heavy objects	Mobile cage and cart handles should be located so as not to expose hands. Crushing injuries from door frames, other adjacent objects; use two persons for moving heavy-wheeled equipment

noninfectious hazards should also involve a qualified health and safety professional with training in ergonomic hazards.

PHYSICAL HAZARDS

Bites and Scratches

Work with nonhuman primates exposes personnel to bite and scratch wounds that not only present zoonotic disease concerns but also can require general medical care, such as suturing and traditional wound management. A wound to the hand may preclude an injured worker from normal occupational activities to allow the wound to heal. Treatment of these wounds is discussed in greater detail in Chapter 9. Prevention of such injuries, which includes cage design, animal handling techniques

TABLE 4-1 Continued

Hazard Type and Operation	Precautionary Measures
Steam and thermal burns. Autoclaves, cage washers, heated pressure washers	Steam and pressure systems must be designed and maintained in accordance with ASME codes; use protective eyewear, face shield, thermal-protective gloves; kerosene-heated or gasoline-powered steam-spray units cannot be operated inside because of potentially toxic carbon monoxide emissions
Noise. Nonhuman-primate holding rooms, other areas where animals are vocal; work near cage washers, other loud mechanical equipment	Conduct sound-level surveys in operations that seem loud; implement a hearing conservation program for all personnel who are exposed at or above 85 dbA (or equivalent daily noise dose); provide hearing protection for high-noise operations; educate personnel on the hazards of noise-induced hearing loss and on the hearing-conservation program
Allergens. Exposure to animal dander, bedding, latex, and other potential allergens.	Minimize exposure to airborne allergens by use of dust-capture ventilation and personal protective dust respirators; remain alert to signs suggestive of an allergic reactions (e.g., rhinitis, conjunctivitis, coughing, sneezing, wheezing)
Heat stress. High-temperature environments, use of PPE that does not allow dissipation of body heat	Provide frequent breaks and drinking water; reduce level and duration of exertion when more protective equipment is worn in high temperatures; consider using personal cooling devices in situations of extreme heat or exertion

and tools, and personal protective equipment (PPE), are also presented in later chapters.

Needles and Other Medical Sharps

Research protocols and general animal husbandry expose workers to various medical sharps that can cause puncture wounds and lacerations. Examples are needles and syringes used for phlebotomy and injection, scalpels and other surgical instruments, dental equipment, and scraping and cutting tools used in cleaning and other daily operations. As with bites and scratches by nonhuman primates, exposures to sharps in a manner that breaks the skin has the potential for not only infection but also injury and disability.

Needle stick injuries are also an important source of exposure to bloodborne pathogens. Revision to OSHA's bloodborne pathogens standard added new requirements for employers to adopt engineering and work practice controls that would eliminate or minimize employee exposure from hazards associated with bloodborne pathogens. As a component of the Exposure Control Plan, employers must take into account new changes in medical technologies and document consideration of commercially available and effective safety devices. This may take the form of an annual review and should list the devices under consideration. The review should describe the devices evaluated for use, the method used to evaluate those devices, and justification for the eventual selection of new devices.

Slips, Trips, and Falls

Slips, trips, and falls are a common cause of traumatic injury to workers in nonhuman-primate facilities. They are not specific to nonhuman-primate facilities; any work setting where floors are wet and there are residues of cleaning agents, feces, or tripping hazards presents a high risk of slip and fall injuries. Personnel carrying heavy loads on wet floors are at high risk of a slip or fall injury. Deteriorated stairs, ladders, and step stools have caused injuries in some facilities. Using slip-resistant footwear helps to reduce the likelihood of such injuries. Guidelines have been established for preventing these injuries (NSC 1997).

Overexertion

Handling and carrying nonhuman primates, heavy equipment, large containers of feed, and other heavy or awkward objects present the risk of musculoskeletal injury. Such injuries commonly are related to a specific and identifiable work task or movement, but they oftentimes occur in people who have pre-existing injuries or are suffering from pathologic conditions. Regardless of a person's medical condition, the risk of injury increases with the weight of the load, the awkwardness of the load and lifting movement, and the number of repetitions. The National Institute for Occupational Safety and Health (NIOSH) has issued a guide for evaluating lifting operations and specifying limits (NSC 1996).

Work activities that are performed repetitively (more than 100 times per day) or for extended periods have an increased risk of causing musculoskeletal disorders. These types of work injuries and illnesses differ from acute traumatic injuries, such as lumbar strain following the lifting of a single heavy object. Repetitive strain injuries and cumulative trauma disorders are characterized by a slower development of impairment and disability and commonly involve a relatively low-force task that workers do not recognize as causing their painful and disabling condition. For example, handling hundreds of cages in the same way every day is an example of the kind of situation in which laboratory animal workers have suffered musculoskeletal disorders.

Hand Injuries from Cage Moving and Cleaning Operations

In nonhuman-primate operations that use portable cages, serious hand injuries have occurred when workers have only the exterior cage framework to grasp when moving cages. Many types of wheeled-cage systems weigh several hundred pounds and lack handles that allow for biomechanically efficient gripping and movement. Gripping the vertical exterior framework commonly exposes workers' hands to crush injuries when they try to move the bulky cages through doorways and in crowded areas. Gripping vertical corner pieces also exposes the hands and forearms to scratches by the nonhuman primates when the cages are occupied. As cages are replaced, new mobile cage units with handles that facilitate safe movement should be obtained. Units that lack such handles or that are not designed to allow safe movement should be modified.

Pressurized-steam and heated-water washers are used to clean nonhuman-primate caging and housing units. Steam washing systems can be portable units or can operate with a centralized steam supply within the facility. Direct skin contact with the steam washing stream or with noninsulated piping or handles can cause burn injuries that can be severe. Eye contact with a direct stream of pressurized steam can cause serious eye injury, including blindness.

Pressurized-steam washing units must be engineered in accordance with American Society of Mechanical Engineers (ASME) codes (ASME 2001) and incorporate features that minimize the potential for operator steam and thermal injuries. Design features should include a control valve that automatically stops pressurized steam flow unless the operator is gripping it. The design of dispensing wands and nozzles should be based on ergonomic principles so that excessive gripping pressure is not required to activate the steam flow or to hold onto the unit. Steam distribution pipes and hoses must be insulated where personnel could touch the surfaces. All components that could be pressurized must be rated for the maximal foreseeable pressure.

PPE that should be worn during use of steam washer units includes:

- Eyeware in accordance with American National Standards Institute Standard Z87 for use under a protective face shield
- Splash-protective outer garments to prevent liquid splashes from reaching the head, skin, and undergarments

- Thermal protective gloves
- Hearing protectors if noise reaches or exceeds 85 dbA per 8-hour time-weighted average (TWA)

Stationary rack washers and autoclaves are commonly used in nonhuman-primate facilities to clean and disinfect mobile cage and housing systems. Burn hazards are controlled by maintaining protective guards, panels, or insulation on the equipment. Thermal protective gloves and arm covers should be used for any operation in which direct skin contact could occur.

In addition to heat and steam, autoclaves present hazards associated with pressure. Personnel who operate or maintain autoclave equipment must be trained, certified, or otherwise specifically qualified for such work, because of the high temperature and pressure hazards. Lockout/tagout work practices, which are focused on preventing electrical, chemical, and/or steam energy from harming employees, are required for many maintenance operations on this equipment.

Kerosene-heated portable units can be used only outside and in well-ventilated areas and if not left unattended for extended periods of time. The products of combustion can include carbon monoxide and other toxic emissions that can create unhealthful air contamination.

Noise

Work in nonhuman-primate housing areas and such activities as cage washing can expose personnel to noise that can cause material and insidious hearing loss over time. Excessive noise may also occur with certain species of nonhuman primates through their vocalization or cage displays. It is recognized that noise-induced hearing loss will develop from workplace exposures when a person is exposed to 85 dbA or more during an 8-hour workday. Operations that seem loud and impair communication between personnel should be evaluated with sound-measurement instruments. Personnel who are exposed to more than 85 dbA per 8-hour TWA must be included in a hearing-conservation program that includes training, use of hearing protectors, and audiometric testing.

Allergens

Personnel working in nonhuman-primate settings must be alert to symptoms and signs of allergic reactions. Inhalation allergens may include airborne animal dander and bedding. Contact allergens causing dermatitis may include gloves and other protective clothing. In one study of laboratory-animal associated allergies, 23.6% of nonhuman-primate handlers reported allergic symptoms (Aoyama and others 1992). Acute asthma has been reported after exposure to nonhuman-primate dander (Petry and others 1985). Exposure to allergenic proteins in latex that is used for protective gloves and other equipment has been well documented (Levy and Leynadier 2001). It should be noted that masks that are suitable for protection against allergens are not the same as those to be worn for protection from splash exposures.

Hazards Associated with Using Personal Protective Equipment

Gloves, protective garments, and eyewear are of great value in forming a protective barrier against some types of hazards, but they can also create hazards themselves. Use of protective equipment in high temperatures must take into account the risk of serious heat-stress injuries and dysfunction. Personnel should be provided with appropriately scheduled break periods, allowing them to replenish their volume losses of water and minimize the effects of reduced cooling ability when PPE is being used under high ambient temperature conditions.

Respiratory protective equipment must be selected on the basis of the specific airborne hazards present, and it must be medically determined that a person is fit to use it.

Protective eyewear can become fogged by perspiration and exhaled breath. Such impairment of vision can create a hazard by decreasing visual acuity and decreasing the user's ability to avoid animal bites and other hazards. Special attention must be given to providing personnel with high-quality protective eyewear that allows them to do their jobs and have appropriate eye protection.

CHEMICAL HAZARDS

Several chemical hazards can be encountered in work with nonhuman primates (Table 4-2), such as disinfectants and volatile anesthetics. It is critical to maintain an inventory of hazardous chemicals to use material safety data sheets, and to train workers in the hazards of and precautions for work with chemicals (NRC 1997; NSC 1997). The potential risks of exposure to hazardous chemicals used in experimental protocols should also be considered. For example, the neurotoxin MPTP (1-methyl-4-phenyl-1,2,3,6 tetrahydropyridine) is commonly used in nonhuman-primate models of Parkinson's disease.

TABLE 4-2 Summary of Chemical Hazards

Chemical Hazard and Operation	Precautionary Measures
Chemical disinfectants and cleaning solutions. Sodium hypochlorite, phenolic solutions, hydrogen peroxide, quartenary ammonium compounds, chlorine dioxide solutions, glutaraldehyde	Prevent splash exposures to corrosive solutions by using face shields or safety goggles; avoid contact by use of impervious gloves, apron, sleeve covers, boots, other chemical-impervious equipment; where potentially injurious gaseous or aerosol air-contaminant exposures are possible, provide capture exhaust ventilation or respiratory protection
Volatile anesthetics. Isofluorane, halothane, nitrous oxide, enflurane	Use a scavenging anesthesia system to capture waste anesthetic gases; vacuum or mechanical venting of scavenging system is most effective; do not use electrocautery or other sparking equipment where oxygen and anesthetic gases are being released

Disinfectants

Disinfectant solutions are used extensively in nonhuman-primate settings and have been the cause of skin and eye irritation and corrosive burn injuries. Disinfectant cleaners that are in common use include:

- Bleach (sodium hypochlorite)
- pH-optimized bleach (pH 7.2)
- Chlorine dioxide solutions
- Quaternary ammonia mixtures
- Phenolic cleaners
- Hydrogen peroxide-peracetic acid cleaners
- Glutaraldehyde and formaldehyde fumigant

All of these cleaners have the potential for causing corrosive injury to the eyes and skin as a result of direct liquid splashes. The choice of protective eyewear and chemical-protective footwear, gloves, and other clothing depends on how the disinfectant is being applied. Application of a disinfectant in accordance with the manufacturer's label with a hand mop can generally be performed safely if the user wears protective eyewear, gloves, boots, and coveralls or an apron that provide a barrier to direct splashes. An industrial hygienist or safety professional should be consulted on safety precautions if fumigation or pressurized sprayer application of chlorine dioxide, pH-optimized bleach, hydrogen peroxide, or aldehyde disinfectants will be conducted. Respiratory protection is commonly specified for such operations.

Volatile Anesthetics

Volatile anesthetic agents are used in nonhuman-primate operations during surgical procedures performed for animal health care and research-protocol purposes, typically including:

- Halothane
- Nitrous oxide
- Enflurane
- Isoflurane

Most of the volatile anesthetic agents are potentially toxic and have effects on the liver and nervous system and evidence of increased risk of adverse reproductive effects. NIOSH has issued guidelines for control of exposure to anesthetic agents in surgical operations (NIOSH 1977). Exposures can be reduced by scavenging vented fumes from rebreathing anesthetic units, using tight-fitting endotracheal tubes and induction masks, and filling vaporizer units carefully so that spillage does not occur. However, administration of these agents to animals can still result in release of anesthetic vapors into the area around the anesthetic equipment and around the animal's mouth and nose, since waste anesthetic gas scavenging systems are generally imperfect.

5
Risk Assessment: Evaluating Risks to Human Health and Safety

BACKGROUND

Risk assessment is a powerful tool that provides a rational framework for designing and managing an OHSP at institutions that use nonhuman primates. The process of risk assessment requires a factual base to define the likelihood of adverse health effects of workplace-associated injuries and exposures, and it attempts to balance scientific knowledge with concerns of staff, investigators, administration, and the public at large. It involves a systematic approach to the identification and characterization of physical, chemical, and biologic hazards to individuals and populations in their environment. The consequences of such hazards can include severe illness or injury, an irreversible health consequence, an unfamiliar disease, and an undesirable situation that might have been avoided by use of an alternative approach or technology. Risk assessments typically require that attention be given first to the most important hazards, that is, the ones that can result in the worst health-related outcomes.

Successful risk assessment offers many advantages. For staff members, a well-defined assessment of risks in the workplace can provide a rational basis for safe practices and behavior. For institutional managers, a well-defined assessment of risks can provide clear targets for injury-prevention and exposure-prevention programs. For regulators and other oversight bodies, a well-defined assessment of risks helps in setting workplace health and safety standards and in monitoring compliance without the need for case-by-case judgments. For concerned citizens, a well-defined assessment of risks provides a concise focus for evaluating protection of the public welfare.

The purpose of risk assessment is to determine the probability of injury or illness due to specific hazards. Risk assessment also includes characterization of the uncertainties inherent in the process of inferring risk. The process in turn becomes the basis of risk management—courses of action to mitigate hazards at the national, regional, and local levels through the establishment and modification of regulatory standards and institutional occupational health and safety programs. Several key terms and concepts are used in risk assessment, including the following (NRC 1983; Osborne and others 1995):

- *Hazard*: A source of risk, such as a substance or action that can cause harm.
- *Exposure*: Contact with a hazard in such a manner that effective transmission of the agent or harmful effects of the agent may occur.
- *Dose-response relationship*: A relationship in which a change in amount, intensity, or duration of exposure is associated with a change in the risk of the outcome.
- *Risk*: The combination of the likelihood (probability) and magnitude (severity) of an adverse event.
- *Uncertainty*: An instance of limited knowledge, false assumption, or statistical variability that contributes to a statement of confidence in conclusions drawn from a risk assessment.
- *Risk management*: The process of formulating and implementing a course of action to mitigate hazards determined by risk assessment to be important.

THE PROCESS OF RISK ASSESSMENT

The process of risk assessment, as used by US regulatory agencies charged with protecting workers and the general public, involves four sequential steps (NRC 1983; Samet and Burke 1998): hazard identification, dose-response assessment, exposure assessment, and risk estimation and characterization. Multiple sources of data may be used to complete each step, including on-site review and investigation, epidemiologic investigation, surveillance, laboratory animal studies, and computer modeling (see Table 5-1). At the institutional level, risk assessments need not be formal endeavors led by recognized experts, but should focus on the same basic steps with most of the emphasis on hazard identification and exposure assessment. Often this will also reveal likely determinants of exposure to the hazards that should be addressed in the institution's occupational health and safety plan.

TABLE 5-1 Assessment of Risk Associated with Animal-Related Research

Criterion	Possible Classifications	Information Sources
Exposure intensity	High Medium Low Absent	Job profile, environmental health and safety assessment, employee history
Exposure frequency	8 h/wk or more Less than 8 h/wk No direct contact	Job profile, environmental health and safety assessment, employee history
Hazards posed by animals	Severe illness Moderate illness Mild illness Illness unlikely	Institutional veterinarian, physician
Hazards posed by materials used in or with animals	Severe illness Moderate illness Mild illness Illness unlikely	Material-safety data sheets; CDC-NIH agent summary statements; radiation-, chemical-, and biological-safety committees; environmental health and safety staff; direct observation, principal investigator
Susceptibility of employee	Direct threat[a] Permanent increase Temporary increase	Medical evaluation, review of personal medical records
Expected incidence	High Medium Low	Published reports, industry experience
History of occupational illness or injury in the position or workplace	Common Uncommon Rare/exceptional	Worker-compensation reports, OSHA 200 log
Regulatory requirements	Required Professional judgment permitted	Environmental health and safety office, consultants, risk managers

[a]Reasonable probability of substantial harm. Americans with Disabilities Act of 1990 (PL 101-336).

SOURCE: Adapted from NRC 1997.

Research involving awake-behaving nonhuman primates requires special consideration during risk assessment. Nonhuman primates can weigh more than human beings, have considerable speed, strength, and manual dexterity, and can harbor zoonotic infectious agents. They can inflict serious physical injury and cause life-threatening illnesses to persons around them. One way to reduce the risks from these animals is by training them to perform certain movements. For example, macaques and squirrel monkeys can be trained to move voluntarily from the home cage into a restraint chair (Ator 1991). Another consideration when working with awake-behaving nonhuman primates is that they are often transported to testing facilities (e.g., laboratories or imaging facilities) outside of the animal quarters. The animals may traverse common use corridors and elevators, potentially exposing individuals not involved in the animal care program. Individuals in other areas of the building may also be exposed if the air exhausted from a testing facility is recycled into other building areas. For these reasons, procedures involving awake-behaving nonhuman primates must undergo additional hazard identification and risk assessment. It is also important that risk assessment of noninfectious hazards involve a qualified health and safety professional with training in the chemical and ergonomic hazards associated with their use. More detailed guidelines for working with awake-behaving NHP are forthcoming from the ILAR Committee on Guidelines for the Care and Use of Mammals in Neuroscience and Behavioral Research (NRC In press). This report will identify common research themes in contemporary neuroscience and behavioral research based on input from neuroscience and behavioral researchers most familiar with current standards of practice and veterinarian specialists in laboratory animal medicine; provide collective, professional judgment in applying current animal care and best use practices to procedures in these areas of research; provide information about new scientific and responsible use developments used to maintain animals during these experiments; and serve as an informational resource to assist researchers, laboratory animal medicine veterinarians and IACUC members in the interpretation and implementation of current standards of practice and promote the training of animal care specialists in this area.

Hazard Identification

The identification of hazards is typically a qualitative process, most often based on observation, experience, published reports and professional judgment. Hazards in the work environment can be identified by safety specialists using institutional logs, worker-compensation reports, and other information sources (NRC 1997) as well as direct observation of the animal facility. That process should be systematic and based on the principles of biologic, chemical, and physical safety; modes of transmission of infectious agents; understanding of the facility design, equipment, personal protection devices, and practices; and knowledge of applicable local, state, and federal regulations. Chapters 3 and 4 of this report provide an overview of the infectious and noninfectious hazards identified in the use of nonhuman primates in research. If the existence of a hazard cannot be definitively shown in the first step in the risk assessment process, the subsequent steps generally are not warranted.

A review of worker exposure and injury reports suggests that most workplace hazards found in nonhuman-primate research facilities are similar to those found in other laboratory animal research environments (bin Zakaria and others 1996; Poole and others 1998, 1999). Most common are animal-inflicted trauma from bites, cuts, and scratches; punctures from needle sticks and other sharps; musculoskeletal injuries, such as strains and sprains, especially involving the back; repetitive-motion injuries (ergonomic injuries); slips, trips, and falls; contact with allergens and chemicals; burns caused by contact with hot surfaces or steam; and various suspected exposures to materials that potentially contain infectious agents.

In addition to a review of worker injury and exposure logs, inspection of the facilities will assist in the identification of hazards. Various aspects of the facility design should be evaluated, such as air-exchange rates, air recirculation, and pressure differentials; use of high-pressure hoses, steam, or other cleaning methods; wastewater drainage; composition and uniformity of ceilings, walls, and floors; and laboratory access requirements and controls. Summaries of prescribed safety measures may be obtained from the IACUC and/or the environmental health and safety office responsible for review and approval of facilities. Inspections of the facility should also consider the daily flow of all materials relevant to potential hazards in the institution, such as the primates themselves and their tissues, caging equipment, environmental-enrichment devices and other husbandry items, animal waste, laboratory waste, cage-wash machinery and supplies, research and veterinary supplies and equipment, and facility-maintenance and janitorial equipment and supplies. For example, soiled primate caging may be moved along common use hallways of the animal quarters to the cage wash area, potentially exposing research staff not associated with the primate program. Each situation should be evaluated systematically with the biologic, chemical, and physical hazards associated with nonhuman primates and in conjunction with objective criteria and information sources.

Dose-Response Assessment

The next step in risk assessment deals with the dose-response relationship. This action establishes the relationship between the quantitative level of a hazard and the probability of an adverse response to it (NRC 1983). Cumulative-dose effects over years of exposure are relevant to chemical toxicants, but single-event exposures are more often of primary concern in the case of pathogenic agents. In both cases, each potential exposure episode is regarded as increasing the cumulative risk, but the likelihood of disease after any single exposure episode is considered to be the same for everyone in the risk group (OSHA 1991). Threshold and non-threshold models of risk exist for various types of health conditions, predominantly with respect to noninfectious hazards, which adds controversy and complication to the risk assessment process (NRC 1983).

Exposure Assessment

Once a hazard has been identified and the dose of the hazard that causes adverse consequences is determined, the next step in risk assessment is exposure assessment. This step estimates the exposure or contact between a hazard and a person (NRC 1983). Exposure assessment must take into account numerous modes of possible contact, such as splashes, bites, aerosols, and needle sticks. The extent to which people are in contact with potential hazards should be determined in conjunction with their job duties and the use of personal protective equipment. Exposure assessment must include evaluation of the experience and skill levels of people who are at risk for exposure. For example, in environments where exposure is associated with failure to comply with standard operating procedures or to use equipment properly (CDC-NIH 1999; NRC 1997), inexperienced personnel would have a greater risk of exposure than more experienced personnel.

Care should be taken when estimating exposure on the basis of injury and exposure logs. Under-reporting of occupational injuries and exposures to supervisors or health-care staff by nonhuman-primate research workers is common. A recent study (bin Zakaria and others 1996) found that 59% of animal-inflicted scratches, 50% of mucous-membrane exposures, and 20% of needle stick injuries went unreported. A variety of reasons were given by respondents in the survey, most commonly that reporting was "too much trouble," that the injury was "not serious enough to report," and that "the injury was accepted as a routine risk." Others have suggested that only 45-68% of injuries are reported to supervisors and that for 4-8% of injuries, workers take no action (Sotir and others 1997). Those observations have important implications for effective communication of the hazards associated with exposure to nonhuman primates and their tissues, especially in light of the finding that occupational infections with B virus have not correlated with injury severity (Hilliard and Henkel 1998).

Risk Estimation and Characterization

The final step in risk assessment is risk estimation and characterization. In this step, the dose-response relationship and exposure assessment are combined to describe the risk to subject persons (NRC 1983). It is essential that persons responsible for conducting risk assessments be knowledgeable about the physical, biologic, and chemical hazards present in nonhuman-primate research, as outlined in Chapters 3 and 4. General principles of safety as they pertain to each hazard should be understood, including essential aspects of the laboratory, husbandry, and veterinary equipment in use; facility design elements, such as the systems for air handling and waste decontamination and disposal; systems of employee hygiene and medical surveillance; and how all these are integrated into the OHSP. Persons responsible for risk assessment must also have an appreciation of the flow of the typical workday activities of animal care, facility maintenance, and research as performed by the different members of the staff (and, if applicable, students and visitors). Knowledge of local, state, and federal regulations under which the facility operates is also important.

As noted throughout this report, the risk assessment process should initially focus on the greatest hazards, those with potential for important consequence for the greatest number of persons. For example, institutions that use macaques or their tissues should first ensure that the possibility of B virus exposure has been assessed. Some of the resulting safety measures will reduce potential injuries from other sources, as in the use of splash barriers, which protect the mucous membranes of the face against infectious agent exposures and chemical exposures during research and husbandry operations.

Once hazard identification is accomplished, other steps in risk assessment are aimed at estimating the risks associated with hazards identified in specific institutional settings. A wide variety of analytic tools are used in these efforts, including qualitative, semiquantitative, and quantitative methods to determine the likelihood of an event in a specified interval and the sources and magnitude of uncertainty and variability in the estimates (Hallenbeck 1993).

Most experts agree that risk assessments should be put into quantitative terms to the greatest extent possible (OSHA 1991). This is especially true when conducting risk assessments for the purposes of establishing regulatory safety standards. Limitations of the data available for use in quantifying the importance of specific hazards contribute to the uncertainty in estimates. Nevertheless, it is desirable in the development of OHSPs to add as much quantitative information to qualitative observations and institutional experience as possible, to arrive at the best possible evaluation of risks posed by specific hazards. It is not realistic to defer the process of risk assessment while waiting for data that may never become available.

Risk to workers is best measured through the use of incidence rate calculations, in which the numerator is the frequency (or number of new occurrences) of an event during a specified period and the denominator is the average size of the group considered at risk for the event:

$$\text{Incidence rate} = \frac{\text{frequency of event (or number of new occurrences)}}{\text{average number of people at risk for event}}$$

For example, the incidence of needle stick accidents among veterinary staff involved in the care of nonhuman primates should use as the denominator the size of the veterinary workforce involved in venipuncture tasks in nonhuman primates at the institution. Incidence rates are useful whether the purpose is to compare trends as new safety-related equipment and policies are established in an institution or to compare the experience of different sites or institutions. Standardizing the average size of the group considered at risk for an event by using full time-equivalents, such as person workdays (pwd), in the calculation of incidence rates allows accurate comparisons. However, determination of the approximate size and nature of the group that should be considered at risk requires well-reasoned efforts. Rapid expansion of knowledge regarding the types and sources of hazards demands that this be a continuing commitment.

Understanding of uncertainty in risk assessment is important in conveying the likelihood of an adverse event or the magnitude of its consequences. Reductions in uncertainty do not change the risks, but they increase the mathematical precision of evaluation. Therefore, a clear understanding of the uncertainties included in risk estimates is essential for policy-making, lest misleading information and ineffective action plans result (Hallenbeck 1993). Data for use in quantifying important risk factors are often sparse, and this, combined with differences between information sources and inherent variability, contributes to uncertainty in estimates. By varying the assumptions that are used, a sensitivity analysis can help evaluate the ramifications of variables in risk estimates and yield better predictions of the impact of various management options.

Risk of Infectious Hazards Associated with Nonhuman Primates

Risk assessment of infectious hazards is particularly important in nonhuman-primate research facilities. As described in Chapter 3, nonhuman primates can harbor zoonotic agents—such as B virus, *Mycobacterium tuberculosis*, SIV, and enteric pathogens—some of which have dire consequences.

In performing a qualitative risk assessment, all risk factors are first identified and explored. For infectious hazards, the risk of becoming infected depends upon the likelihood of a relevant exposure to a source of infection and the likelihood of becoming infected if there is an exposure. The following elements may be evaluated when assessing the risk of infectious hazards in the conduct of laboratory animal work: animal contact, exposure intensity, exposure frequency, physical and biologic hazards present by the animals, hazardous properties of agents used in research protocols, susceptibility of employees, and occupational-health history of employees doing similar work (NRC 1997). Exposure intensity measures the estimated dose received among those exposed over some arbitrarily defined unit of time, whereas exposure frequency concerns the number of opportunities for any degree of exposure during the same period. For zoonotic diseases, both of there parameters are affected by the prevalence of the agent in the animals, its shedding pattern, environmental stability, and routes of transmission to humans. Exposure intensity values are often used when setting allowed safety standards to chemical or allergen hazards and are generally more applicable in those cases. Consideration of the importance of individual hazards identified in the workplace should include the size of the group at risk, the potential effects of the hazards, and the magnitude of the exposures (NRC 1997). Ranking of hazards based on their importance can include institutional experience regarding worker illness and injury rates, near-miss reports, reference information, and other documents (see Table 5-1).

As stated in the federal guidelines outlined in *Biosafety in Microbiological and Biomedical Laboratories* (CDC-NIH 1999), the primary role of risk assessment is to aid in the prevention of workplace-acquired infections, and the secondary role is to aid in the prevention of infections in the surrounding community (CDC-NIH 1999). In this context, risk assessment has led to the assignment of designated animal biosafety levels (ABSLs) 1-4 for experimental research activities with specified pathogens. ABSLs are described in terms of facilities, equipment, and practices, each being important in mitigating hazards or risks to workers and the public (see Table 5-2). This approach has a laudable record of contributing to overall workplace safety, despite the difficulty in assessing some variables, such as emerging infectious agents and genomic manipulations. A conservative approach is generally recommended when a lack of information forces subjective decision-making. For example, when infectious-disease risks are being considered, universal precautions are always advisable (CDC-NIH 1999).

However, these federal guidelines do not address noninfectious laboratory hazards, nor were they intended to guide safety considerations outside the laboratory per se, such as in outdoor holding enclosures for nonhuman primates or in settings involving their wild capture and transportation. Thus, OHSPs for persons working with nonhuman primates in many situations must be based on general workplace safety considerations and analogous hazards in other industries without the benefit of the specific algorithms found in the national laboratory biosafety guidelines.

Important characteristics of most well-known infectious agents that are useful in risk assessment are readily found in Chapter 3 and the references cited at the end of this report. Such information is usually based on medical surveillance and epidemiologic studies and grounded in laboratory investigations of the agents themselves. Many agents known to have caused laboratory-acquired infections are listed in Section VII of *Biosafety in Microbiological and Biomedical Laboratories* (CDC-NIH 1999). The following characteristics predict risk, but the characteristics considered as a whole are more important than any one of them individually.

- Pathogenicity
- Stability in the environment
- Infectious dose
- Concentration in specimens or in the environment
- Origin (host, geographic location, or type of source)
- Route of transmission
- Availability of data from animal studies
- Availability of effective prophylaxis or therapy
- Experience and skill of personnel at risk

Selection of the appropriate ABSL for activities involving infectious material in nonhuman-primate research should be based on evaluation of these criteria, with modifications as needed in light of current scientific information. Answers to questions about the characteristics listed above often are not definitive, especially for newly described infectious agents and materials that contain recombinant DNA; in such cases, the risk-assessment process should include an institutional biosafety committee (NIH 1998).

Risk assessment leading to the requirement for an ABSL-2, 3, or 4 presupposes that the workforce is composed of immunocompetent

TABLE 5-2 Summary of Recommended Animal Biosafety Levels (ABSLs) for Activities in Which Experimentally or Naturally Infected Vertebrate Animals Are Used

ABSL	Agents	Practices	Equipment (Primary Barriers)	Facilities (Secondary Barriers)
1	Not known to cause disease consistently in healthy human adults	Standard animal care and management practices, including appropriate medical surveillance programs	As required for normal care of each species	Standard animal facility: • No recirculation of exhaust air • Directional air flow recommended • Handwashing sink recommended
2	Associated with human disease; hazard: percutaneous exposure, ingestion, mucous membrane exposure	ABSL-1 practices plus • Limited access • Biohazard warning signs • Sharps precautions • Biosafety manual • Decontamination of all infectious wastes and of animal cages before washing	ABSL-1 equipment plus • Primary barriers • Containment equipment appropriate for animal species • PPE: laboratory coats, gloves, face and respiratory protection as needed	ABSL-1 facility plus • Autoclave available • Handwashing sink available in animal room • Mechanical cage washer used

3	Indigenous or exotic agents with potential for aerosol transmission; disease can have serious health effects	ABSL-2 practices plus • Controlled access • Decontamination of clothing before laundering • Decontamination of cages before bedding removed • Disinfectant foot bath as needed	ABSL-2 equipment plus • Containment equipment for housing animals and cage-dumping activities • Class I or II BSCs available for manipulative procedures (inoculation, necropsy) that might create infectious aerosols • PPE: appropriate respiratory protection	ABSL-2 facility plus • Physical separation from access corridors • Self-closing, double-door access • Sealed penetrations • Sealed windows • Autoclave available in facility
4	Dangerous or exotic agents that pose high risk of aerosol transmitted infection or life-threatening disease; or related agents with unknown potential for transmission.	ABSL-3 practices plus • Entrance through change room where personal clothing is removed and laboratory clothing is put on; showering on exit • Decontamination of all wastes before removal from the facility	ABSL-3 equipment plus • Maximal containment equipment (Class III BSC or partial containment equipment in combination with full-body, air-supplied positive-pressure personnel suit) used for all procedures and activities	ABSL-3 facility plus • Separate building or isolated zone • Dedicated supply and exhaust, vacuum and decontamination systems • Other requirements outlined in the text

SOURCE: Reprinted from CDC-NIH, 1999.
NOTE: ABSL – animal biosafety level; BSC – biological safety cabinet; PPE – personal protective equipment.

people. Immunocompromised people are at increased risk in many cases when exposed to infectious agents. That is just one of the complexities that can enter into risk assessment, so other variables, as described in Section VII of *Biosafety in Microbiological and Biomedical Laboratories* (CDC-NIH 1999), must also be considered in evaluating risks to particular workers. In all research involving the use of nonhuman primates, the study director or principal investigator must work with the IACUC, the biosafety officer, and the primate center director to assess risks and set ABSLs in the context of the institutional administrative structure; ultimate authority rests with the senior institutional official (CDC-NIH 1999).

RISK OF OCCUPATIONAL INJURIES AND EXPOSURES AT NATIONAL PRIMATE RESEARCH CENTERS

Physical Hazard Risk Assessment

An epidemiologic investigation of work-related injuries and exposures among animal care, veterinary, and scientific staff at a US regional primate research center provided yearly estimates of incidence rates ranging from 44 to 65 animal-associated injuries per 100,000 pwd during the 5-year period of observation (bin Zakaria and others 1996). Animal-inflicted scratches and bites had the highest 5-year incidence rates (82.1 and 80.8 incidents per 100,000 pwd, respectively), together accounting for 51.7% of reported incidents. Cuts and mucous membrane exposures had the lowest 5-year incidence (45.0 and 17.6 incidents per 100,000 pwd, respectively). Fingers and thumbs were the most common anatomic sites of occupational bite injuries, and full-time workers were 3-4 times more likely to report injury episodes than part-time workers (those with less than 20 hours of animal contact per week). The injury-specific incidence rates differed with job category; veterinary residents in training had the highest overall injury rates. The frequency of all injury types decreased with increasing years of employment, and 33% of all reported injuries occurred in persons hired less than 6 months previously. Those findings have implications for risk assessment at individual institutions, where a workforce composed of many inexperienced people should be assessed at greater risk than institutions with an experienced workforce. The findings also offer an opportunity to compare some work-related injuries incurred during nonhuman-primate handling (such as needle sticks and mucous-membrane exposures) with injuries incurred by persons employed in human hospitals and other health-care settings, which could lead to improvements in safety-training and injury-prevention programs.

B Virus

A quantitative risk-assessment study of primate-associated injuries and exposures at another US national primate research center has been reported (Weigler and Ponce 1999); it was based on analysis of institutional bite-scratch-splash exposure records over a 5-year period. In this case, a stochastic (random) simulation model was done to estimate the efficiencies of different B virus exposure prevention methods. Simulations were done using 2000 iterations with median latin hypercube sampling (Analytica Software, Decisioneering, Denver, CO) assuming that sources of risk were independent and that there was no threshold for exposure. The probability of B virus exposure among workers was estimated by including separate distributions for the prevalence of B virus among macaques in this setting (modeled as a normal distribution with a mean of 0.5 and standard deviation of 0.1) and the likelihood of shedding among infected animals (modeled as a triangular distribution with a mean of 0.02 and range 0 to 0.05). Each category of prevention method (protective eyewear, gloves, mask, and labcoat, laboratory procedures, and postexposure scrub) was included as a separate model for each type of exposure (bite/scratch, needle stick, cut, mucous membrane, and other), using two different statistical distributions (triangular and beta) for exploration. Empirical reasoning was used for parameter estimates of those distributions; for example, protective eyewear was given a mean of 0.75 and range 0 to 1 for protecting mucous membranes but no protection against bites, scratches, or other types of exposure. The actual institutional injury exposure record data were annualized for 8-hour person workdays at risk and stratified by type of worker (veterinary, husbandry, research, maintenance, student, other).The result of these simulation models was the expected incidence of B virus infections for the at-risk population of workers in the institution. That approach led to the prediction of one new human B virus exposure episode per 60 years in the institution, assuming a fixed population size of workers at risk. The study also included a sensitivity analysis of model predictions of the potential impact of different B virus risk-management strategies. Use of PPE that reduced scratch rates, improved laboratory procedures, and increased postexposure wound disinfecting efficiency was most influential in reducing the risk of B virus among workers in these models.

Simian Immunodeficiency Virus (SIV)

In a study involving a questionnaire-based survey of 550 persons working at 13 North American research institutions (Sotir and others 1997), a high frequency of needle sticks and mucocutaneous exposures (defined as animal-inflicted bites and scratches) was documented among persons working with nonhuman primates and their tissues. Over one-third of study participants were reported to have experienced needle sticks or mucocutaneous exposures while working with nonhuman primates, predominantly macaques but including at least six other genera. The study included serial serologic testing for SIV antibodies among study participants and considered whether there was exposure to SIV in the laboratory or to SIV-infected animals. Statistical methods were used to assess possible associations between workers' job categories, job tasks performed, length of employment, work with HIV-2 and SIV, work with nonhuman primates, and the frequency and types of injuries sustained in the workplace.

Persons working with monkeys that were SIV-negative or whose SIV status was unknown were more likely to have sustained (or to have reported) needle sticks or mucocutaneous exposure involving blood, body fluid, or unfixed tissue than were those working with SIV-infected animals. Those study results suggested that increased awareness led to improved safety practices or alternatively to different reporting rates. Some injury-specific frequencies differed with job category, but in contrast with the previously discussed study (bin Zakaria and others 1996), increasing years of employment increased the likelihood of injury occurrence.

Analysis of survey responses showed that persons responsible for more invasive tasks with animals (such as phlebotomy, dental work, surgery, necropsy, and experimental inoculation) were at greater risk for needle stick injuries than persons doing noninvasive work (such as husbandry, sanitation, and routine medication) even when years of experience were taken into account. However, bite and scratch rates did not differ with task type. Those findings parallel observations of health-care workers at risk for bloodborne pathogen exposure in occupational settings and highlight opportunities for focused preventive educational programs, especially for some occupational groups (such as husbandry staff) that might be less informed about work-related hazards (OSHA 1999).

6
Occupational Health and Safety Regulations and Recommendations Applicable to Nonhuman-Primate Research Facilities

Once a risk assessment has been completed, the final step before developing a risk management plan pertaining to OHS is understanding pertinent OHS standards. Leadership at institutions where nonhuman primates are used in research, teaching, or testing not only must be aware of workplace hazards and associated risks when developing their OHSP but also must be knowledgeable about and compliant with applicable regulations and guidelines. Those prerequisites can be a challenge in that multiple agencies or regulations may be applicable in a given institution. Regulation may be mandated at the federal and state levels and may depend on whether the nonhuman-primate work in question occurs in a federal facility or a federally funded institution. This chapter describes the important regulations and guidelines and provides contact information on organizations that can provide guidance and education on safety standards. Much of the information in this chapter can be found in *Guide for the Care and Use of Laboratory Animals* (NRC 1996) and *Occupational Health and Safety in the Care and Use of Research Animals* (NRC 1997) and is restated here for ease of use.

FEDERAL OCCUPATIONAL HEALTH AND SAFETY REQUIREMENTS

Occupational Safety and Health Act

The Occupational Safety and Health Act of 1970 mandates that all nongovernment employers provide a safe and healthful workplace for their employees. It also provided for the creation of the Occupational Safety and Health Administration (OSHA) and the National Institute for Occupational Safety and Health (NIOSH). The act directs OSHA to develop and issue standards through a public rule-making process. Employers must comply with those OHS standards as they would with any statutory requirement (Blosser 1992).

The most important federal standards governing OHS are found in OSHA standard 29 CFR Part 1910 (www.osha.gov; Standards 29CFR). That document contains many sections that are pertinent to the work conducted in facilities where nonhuman primates are used in research, teaching, and testing. For example, the OSHA bloodborne pathogens standard (29 CFR 1910.1030) requires institutions to provide hepatitis B vaccinations to employees who handle blood, organs, or other tissues from experimental animals infected with hepatitis B virus and to make a confidential medical evaluation available to the employee immediately after an exposure to animal tissues that are contaminated with a bloodborne pathogen. The OSHA standard on occupational exposure to hazardous chemicals in laboratories (29 CFR 1910.1450) requires medical surveillance when monitoring reveals an exposure that routinely exceeds the action level for an OSHA-regulated substance, such as a time-weighted average of 0.75 ppm or a short-term exposure level of 2.0 ppm for formaldehyde (29 CFR 1910.1048). The OSHA standard regulating the use of compressed gases is 29 CFR 1910.101, personal protective equipment (PPE) standards are in 29 CFR 1910.132-1910.140, and electric systems requirements are in 29 CFR 1910. 301-1910.330 although other sections of 29 CFR 1910 might also be applicable.

It is important to note that the OSHA standards are not all-inclusive. They do not directly address every hazard or risk present at every worksite. To address hazards not covered by a particular standard, OSHA may cite Section 5(a)(1) of the Occupational Safety and Health Act (also designated 29 USC § 654(a)(1)); this provision is called the general duty clause because it imposes on employers the general obligation of furnishing workplaces that are "free from recognized hazards that are causing or are likely to cause death or serious physical harm" (Blosser 1992).

Occupational Safety and Health Administration

OSHA (www.osha.gov) is the federal agency charged with protecting the health of employees and preventing occupational injuries, disease, and death. Although OSHA and NIOSH were created by the same act of Congress, they are distinct agencies with separate responsibilities. OSHA is responsible for creating and enforcing workplace safety and health regulations. It establishes protective occupational standards and enforces them through inspection and monetary penalties. It also provides free on-site support to identify and correct hazards and provides assistance in setting up OHS programs.

National Institute for Occupational Safety and Health (NIOSH)

NIOSH (www.cdc.gov/niosh/homepage.html) is the federal agency responsible for conducting research on and making recommendations for the prevention of work-related disease and injuries. NIOSH and OSHA often work together toward the goal of protecting worker safety and health.

NIOSH can be an excellent source of information on OHS for institutions and individuals. It publishes hazard-specific guidance ("hazard IDs") such as *Cercopithecine herpesvirus 1 (B Virus) Infection Resulting from Ocular Exposure,* which is available online at www.cdc.gov/niosh/ and provides a summary of key points, a description of the hazard, recommendations for preventing B virus infections, recommended actions, and references for additional information.

Occupational Health and Safety in Federal Facilities

The Occupational Safety and Health Act in Section 19, also designated 29 USC 668, requires heads of federal agencies to establish and maintain comprehensive and effective OHS programs consistent with the standards set for nongovernment employers by OSHA (such as 29 CFR 1910), although no inspection oversight is detailed. Presidential Executive Order 12196, issued in 1980, further defines the responsibilities of federal agencies, including inspection requirements for federal OHS programs. The inspection requirements for federal agencies are listed in 29 CFR 1960; for example, inspectors must be safety or occupational health specialists or other persons with sufficient training or experience to recognize hazards and suggest general abatement procedures (29 CFR 1960.28(a)); and all agency workplaces must be inspected annually, and more frequent inspections must be conducted in workplaces where there is an increased risk of accident, injury, or illness (29 CFR 1960.25(c)). Furthermore, responsibility for oversight of federal OHS programs is assigned to the Office of Federal Agency Programs (OFAP), which functions as a "mini-OSHA" for federal employees. In essence, federal agencies must comply with the same OSHA standards as nongovernment employers; however, they are inspected by OFAP rather than OSHA, and they are not subject to monetary penalties.

Occupational Health and Safety Requirements for Federally Funded Institutions

The *Public Health Service Policy on Humane Care and Use of Laboratory Animals* (OLAW 2000) requires institutions that receive federal funds to have an OHS program as part of their overall animal care and use program. PHS Policy requires institutions to use the *Guide for the Care and Use of Laboratory Animals* (NRC 1996) as a basis of an institutional program for activities involving animals. The *Guide* includes OHS guidelines related to hazard identification and risk assessment; personnel training; personal hygiene; facilities, procedures, and monitoring; personal protection; and medical evaluation and preventive medicine for personnel. The *Guide* is discussed further later in this chapter.

The Office of Laboratory Animal Welfare (OLAW), in the Office of Extramural Research at the National Institutes of Health (www.grants.nih.gov/grants/olaw/olaw.htm), plays an important role in the implementation of PHS Policy. OLAW exercises compliance oversight of PHS Policy on research conducted or supported by any component of the PHS through approval of Animal Welfare Assurances. Each institution that receives federal funding must provide a written Assurance to OLAW. The Assurance Statement outlines how the institution will comply with PHS Policy, including how it will comply with the OHS guidelines outlined in the *Guide*. Once an Assurance Statement is accepted by OLAW, the institution is considered an "Assured" program. OLAW encourages programs to observe applicable industry standards, such as the National Research Council's *Occupational Health and Safety in the Care and Use of Research Animals* (NRC 1997).

OLAW also requires the IACUCs of Assured programs to conduct semiannual program reviews and provides a suggested checklist to assist IACUCs in reviews. The OHS portion of the checklist, which is available online, is as follows (www.grants.nih.gov/grants/olaw/sampledoc/chek1a.htm):

1. Institutional program for a safe and healthy workplace
    - program is established and implemented
    - covers all personnel who work in laboratory animal facilities
    - based on hazard identification and risk assessment
    - personnel training (e.g., zoonoses, hazards, pregnancy/illness/immunosuppression precautions)
    - personal hygiene procedures (e.g., work clothing, eating/drinking/smoking policies)
    - procedures for use, storage & disposal of hazardous biologic, chemical, and physical agents
    - specific procedures for personnel protection (e.g., shower/change facilities, injury prevention

2. Program for medical evaluation and preventive medicine for personnel
   - pre-placement evaluation including health history
   - immunizations (e.g,. rabies and tetanus) and tests as appropriate
   - zoonosis surveillance as appropriate (e.g., Q-fever, tularemia, Hantavirus, plague)
   - procedures for reporting and treating injuries, including bites, etc.
3. Special precautions for personnel who work with primates
   - tuberculosis screening includes all exposed personnel
   - training and implementation of procedures for bites & scratches
   - education regarding Cercopithecine herpesvirus 1 (B virus)

AAALAC International in its review of accredited institutions found the most common deficiencies in the OHSP were: 1) they were not based on hazard identification/risk assessment, and 2) there was inadequate personal protection and hygiene. Less common deficiencies were found in: 1) personnel training, 2) facilities, procedures, and monitoring, and 3) medical evaluation/preventive medicine (DeLong and others 2001). Aspects of the OHSP that are considered by AAALAC International in assessments of animal care and use programs can be found in the Program Description at HtmlResAnchor www.aaalac.org/download.htm.

Federal Nonhuman-Primate Import and Quarantine Requirements and Worker Protection Recommendations

The provisions of 42 CFR 71 seek to prevent the introduction, transmission, and spread of communicable disease from foreign countries into the states or possessions of the United States. Section 42 CFR 71.53 specifically addresses the importation of live nonhuman primates and restricts that activity to registered importers. Live nonhuman primates may be imported into the United States and sold, resold, or otherwise distributed only for bona fide scientific, educational, or exhibition purposes.

The Division of Global Migration and Quarantine in the National Center for Infectious Diseases at the Centers for Disease Control and Prevention (www.cdc.gov/ncidod/dq/) is responsible for preventing the importation and spread of zoonotic illness capable of causing serious outbreaks of communicable disease in humans (such as Marburg/Ebola, monkeypox, yellow fever, and tuberculosis). Its emphasis is on minimizing exposure to imported nonhuman primates during transit and during the mandatory 31-day quarantine period and vigilant surveillance for zoonotic illness.

Nonhuman-primate importers must register with the CDC and certify that the nonhuman primates imported will be used for science, education, or exhibition as defined in the regulations. They must also implement disease-control measures and isolate the nonhuman primates for 31 days. The importers must report suspected zoonotic illness in the nonhuman primates or in workers and maintain records regarding nonhuman-primate distribution. As of this writing, the CDC program has 28 registered nonhuman-primate importers.

CDC's import-quarantine program activities include development of recommendations for disease-control measures, inspection of quarantine facilities, monitoring of arriving shipments, assessment of disease-control measures, review of animal health records, and investigation of illness reports. Worker-protection measures advocated by CDC for activities in the importation and quarantine of nonhuman primates include:

1. Limiting access to imported animals and tissues during transit and quarantine
2. Implementation of an employee OHS program:
   a. Zoonotic-disease risk and prevention training
   b. Tuberculosis skin-testing
   c. Respiratory protection program:
      - Medical fitness
      - Fit-testing
      - Training
   d. Use of appropriate PPE
3. Worker illness and injury surveillance

Additional requirements and recommendations for registered nonhuman-primate import and quarantine activities pertain to facility design and operation, incorporation of disease-control measures into all standard operating procedures that present risk, engineering controls, waste-handling precautions, and PPE recommendations for all activities during quarantine beginning with the animals' entrance onto an aircraft.

Requirements Related to Import and Export of Nonhuman-Primate Material

National Center for Import and Export

The National Center For Import and Export (NCIE) of the US Department of Agriculture Animal and Plant Health Inspection Service (APHIS) has regulatory authority over the importation of human or nonhuman-primate material that is produced in tissue culture or is a potential or actual zoonotic pathogen. For more information regarding import and export requirements of APHIS, see www.aphis.usda.gov.

US Public Health Service

The US Public Health Service (USPHS) has jurisdiction over all human and nonhuman-primate materials (USPHS 42 CFR - Part 71 Foreign Quarantine. Part 71.54 Etiologic agents, hosts, and vectors). Packages containing etiologic agents or vectors originating in foreign locations must have an importation permit issued by the United States Public Health Service. USPHS can be contacted at: Department of Health and Human Services, Centers for Disease Control, Office of Biosafety, Atlanta, GA 30333, or by telephone at 404-639-3883.

US Department of Transportation

The Department of Transportation (DOT) has regulations regarding the transportation of hazardous materials, which include infected live animals or tissues. DOT can be reached at www.dot.gov; questions can be sent via e-mail to DOT information specialists at dot.comments@ost.dot.gov.

STATE OCCUPATIONAL HEALTH AND SAFETY REQUIREMENTS

The Occupational Safety and Health Act allows states to establish their own programs for issuing and enforcing OHS standards. These state programs are subject to certification by OSHA. States may also, subject to OSHA approval, assert jurisdiction over health and safety issues for which OSHA has no federal standard. State standards must be at least as stringent as the OSHA standards. When OSHA adopts a new standard, the state programs must issue corresponding rules. As of this writing, the following states (and Puerto Rico and the Virgin Islands) have chosen to administer their own OSHA-approved programs:

Alaska	Iowa	New Mexico	Vermont
Arizona	Kentucky	North Carolina	Virginia
California	Maryland	Oregon	Washington
Connecticut	Michigan	South Carolina	Wyoming
Hawaii	Minnesota	Tennessee	
Indiana	Nevada	Utah	

USEFUL REFERENCES

Guide for the Care and Use of Laboratory Animals

As described above, the *Guide for the Care and Use of Laboratory Animals* (NRC 1996) is prescribed by the PHS policy as the basis for institutional animal care and use programs, including OHS programs. Chapter 1 of the *Guide*, "Institutional Policies and Responsibilities," provides substantive guidance on OHS. The *Guide* emphasizes that an effective program must rely on strong administrative support and interactions among several institutional programs, including the research program (as represented by the investigator), the animal care and use program (as represented by the veterinarian and the IACUC), the environmental health and safety program, occupational health services, and administration (for example, human resources, finance, and facility maintenance). Day-to-day safety in the workplace is the responsibility of the laboratory or facility supervisor (such as, principal investigator, facility director, or laboratory animal veterinarian) and depends on maintenance of safe equipment and facilities as well as performance of safe work practices by all employees. The *Guide* also provides guidance on hazard identification and risk assessment; personnel training; personal hygiene; facilities, procedures, and monitoring; animal experimentation involving hazards; personal protection; and medical evaluation and preventive medicine for personnel.[1]

Occupational Health and Safety in the Care and Use of Research Animals

Occupational Health and Safety in the Care and Use of Research Animals (NRC 1997) provides guidelines for the occupational health and safety of institutional employees, visitors, and students who might be exposed to hazards in the course of their work with research animals. The Office for Laboratory Animal Welfare strongly encourages institutions to observe the standards set out in this report, which is also available online.[2] The general concepts set forth apply to many categories of institutions: academic, industrial, and government research institutions; biomedical and agricultural research institutions; and medical and veterinary educational institutions. The report provides the following specific recommendations:

---

[1]Free copies of the report can be ordered at www.nationalacademies.org/ilar. The *Guide* is also available on line at www.nap.edu/readingroom/books/labrats/.

[2]www.books.nap.edu/books/0309052998/html/R1.html#pagetop.

- We recommend that every institution initiate a concerted effort to address the health and safety hazards and the risks of occupational illness and injury that are associated with the care and use of research animals and broaden its occupational health and safety program as necessary to reduce the risks to an acceptable level.
- We recommend that the senior official of an institution demonstrate personal commitment to a safe and healthful workplace, delegate clearly defined duties to those with authority to commit and direct institutional resources, and establish mechanisms for monitoring the success of the occupational health and safety program.
- We recommend that every institution develop a multidisciplinary approach to occupational health and safety that permits the continuing evaluation of potential workplace hazards and of the risks to employees working with animals.
- We recommend that the determination of need for health-care services be based on the nature of the hazards associated with the care and use of research animals and the intensity and frequency of employee exposure to these hazards.
- We do not recommend serum collection and storage as standard components of an occupational health and safety program. They have value only for employees who have substantial likelihood of occupationally acquired infection with an agent that can be monitored serologically.
- We do not recommend a physical examination as the principal surveillance tool for periodic health evaluations. We recommend that a careful history based on a knowledge of workplace risks be used for this purpose.

Biosafety in Microbiological and Biomedical Laboratories

Biosafety in Microbiological and Biomedical Laboratories (CDC-NIH 1999) is a major resource for guidelines on safe handling of infected animals as well as for nonhuman-primate cells and tissues. It includes detailed descriptions of criteria for animal biosafety, infectious agents and biosafety cabinets. This document can be accessed online or ordered from the Government Printing Office.[3]

[3]Online address: www.cdc.gov/od/ohs/biosfty/bmbl4/bmbl4toc.htm; Government Printing Office address: Superintendent of Documents, U.S. GPO, Washington, DC 20402 or online at https://orders.access.gpo.gov/su_docs/sale/prf/prf.html.

Association for Assessment and Accreditation of Laboratory Animal Care International

The Association for Assessment and Accreditation of Laboratory Animal Care International (AAALAC International) is a private, nonprofit organization that promotes the humane treatment of research animals through a voluntary accreditation program. This confidential peer review assesses the quality of all aspects of an animal care and use program, including animal husbandry, veterinary care, institutional policies, and the facilities where animals are housed and used. More specifically, AAALAC International carefully reviews OHS programs and assesses their design, scope, and effectiveness in light of the nature of the animal research being conducted. AAALAC International's standards are based on each country's regulations, the principles outlined in the *Guide for the Care and Use of Laboratory Animals* (NRC 1996), and other broadly accepted reference resources. AAALAC International can be reached online at www.aaalac.org or by sending an e-mail to *accredit@aaalac.org*.

American National Standards Institute

The American National Standards Institute (ANSI) is a private, nonprofit organization that administers and coordinates the US voluntary standardization and conformity assessment systems. The Institute's mission is "to enhance both global competitiveness of US business and the US quality of life by promoting and facilitating voluntary consensus standards and conformity assessment systems, and safeguarding their integrity."

Of particular interest to managers of facilities that use nonhuman primates in research, teaching, and education are ANSI standards Z358.1-1998, "Emergency Eyewash and Shower Equipment," and Z87.1-1989 (R1998), "Occupational and Educational Eye and Face Protection." ANSI can be reached on line at HtmlResAnchor www.ansi.org.

ASTM International

ASTM International (formerly the American Society for Testing and Materials) was organized in 1898 and is one of the largest voluntary standards development organizations in the world. The organization's mission is "to be the foremost developer and provider of voluntary consensus standards, related technical information, and services having internationally recognized quality and applicability that promote public health and safety, and the overall quality of life; contribute to the reliability of materials, products, systems, and services; and facilitate national, regional, and international commerce." Familiarity with ASTM International fluid resistance and permeability standards is beneficial when one is researching and selecting personal protective clothing. ASTM can be reached on line at HtmlResAnchor www.astm.org/.

American Conference of Governmental Industrial Hygienists

The American Conference of Governmental Industrial Hygienists (ACGIH®) is a member-based organization that "advances worker health and safety through education and the development and dissemination of scientific and technical knowledge." ACGIH® is one of the industry's leading publication resources, with over 400 titles related to occupational and environmental health and safety, including the Threshold Limit Values (TLVs®) and Biological Exposure Indices (BEIs®). ACGIH® can be reached on line at www.acgih.org.

International Air Transport Association

International Air Transport Association (IATA) partners are airline suppliers and service providers who participate through partnership programs that provide a forum through which these companies develop industry solutions. The IATA Live Animals Regulations are the recognized worldwide standards for transporting live animals by commercial airlines. Countries, such as the member states of the European Union, enforce the IATA regulations for the transportation of live animals. Government agencies, such as the U.S. Fish and Wildlife Service, and the management authorities of the Convention on International Trade in Endangered Species also enforce the regulations for the packaging of endangered species for international transport. It is essential that shipping, acceptance, and handling staff as well as all others involved in the transportation of live animals be familiar with IATA. IATA can be reached on line at www.iata.org/cargo/index.htm.

7
Risk Management: The Principles Underlying the Design and Implementation of an Occupational Health and Safety Plan

Risk management is the process of formulating and implementing a course of action to mitigate the hazards determined in the risk-assessment process to be important (NRC 1983). The identification of the hazards is discussed in Chapters 3 and 4, and the process of determining the risk associated with them (risk assessment) is discussed in Chapter 5.

Recognized hazards can be managed with a variety of adjustments in work practices, equipment, and facilities. In some cases, key modifications focus on engineering controls (facilities and equipment), in others on administrative changes (such as delegation of decision-making authority to the right level or revision of established safety procedures), and in still others on adoption of new safety-related devices, protective equipment, or research methods. Training programs must be adjusted in concert with these changes to ensure their effectiveness. Clearly, different people must be involved to achieve appropriate advances in the various elements that contribute to improved worker health and safety. Basic and applied research may also be needed to identify, evaluate, and develop the means to deal with specific new hazards and to ensure their practicality and usability in the workplace (Samet and Burke 1998).

It is important to recognize that many factors influence risk management. Public values, politics, economics, legal issues, and technical concerns can all influence the risk-management process locally (as in adjusting standard operating procedures) or nationally (as in adjusting guidelines and regulations). In some cases, external influences force overconservative risk-management decisions and actions; more often, these influences, especially fiscal constraints, lead to less than optimal risk-management decisions and actions.

To be effective, risk management must have two elements: a specific occupational health and safety plan and an appropriate safety culture and working environment. Safety culture is often taken for granted, although it is critical in building an effective risk-management system and a healthful overall work environment. At a basic level, safety culture is the way the institutional administration and workers in an organization feel about risk; feelings, attitudes, and perceptions about risk will influence how it is managed. The safety culture sets the tone of an organization, influencing the consciousness of its people as they conduct their daily activities. The safety culture encompasses an organization's tolerance of risk in its daily operating activities and decision-making processes. The greater the degree to which the administration recognizes the need for effective risk-management in the organization, the greater will be its commitment to the establishment of standards and protocols for identifying, assessing, and managing risks, and the more beneficial the risk-management program will be.

This chapter deals with the foundation of risk management—successful OHSP and possible solutions specifically applicable to work environments involving nonhuman primates are described.

While there are no fundamental differences in the OHSP based on the size of a facility, there are some key differences in developing an OHSP for large institutions, such as a primate center, versus a small vivarium that may have only limited numbers of nonhuman primates. The critical differences are likely to be in the inability of smaller institutions to allocate resources and personnel to the OHSP that may be available at larger institutions specializing in nonhuman primates. These limitations can be addressed in part by obtaining the commitment of the institutional official prior to acquiring the animals and presenting a plan to identify the resources that will be required to properly work with the species in question. The allocation of adequate resources also depends on the oversight of the IACUC, which has responsibility for review of occupational health in the vivarium. Housing of nonhuman primates in a conventional research vivarium may require facility modifications and renovations. In some cases, depending on the species in question, it may be possible to address occupational health and safety concerns by use of appropriate personal protective equipment and modifications to existing standard operating procedures (SOPs). All facilities that plan to house nonhuman primates should identify the specific requirements for nonhuman-primate husbandry and incorporate these features into the design and construction of the facility.

The housing of small numbers of primates requires specialized training for employees (relative to the existing training program). Nonhuman primates require a controlled-access space with rigorous attention to SOPs for safe handling and husbandry. In a facility dedicated to nonhuman primates all employees are part of the "culture" of working with primates, while in a traditional vivarium nonhuman primates represent unusual species that may cause increased levels of curiosity among vivarium workers and affiliated research staff. The establishment of strict SOPs and mechanisms of controlled access are critical to reducing this problem.

An occupational health and safety plan must be developed with consideration of the specific nonhuman-primate species in use due to differences in the size and strength of the animals, the special husbandry practices required, and the risk of human exposure to zoonotic infectious agents. These factors all vary with the nonhuman-primate species in use. These considerations ultimately influence the resources required to maintain nonhuman primates and the necessary elements of the occupational health and safety plan.

At the end of this chapter are checklists (Tables 7-1 through 7-11) to determine if the essential elements of an occupational health and safety plan are being addressed at an institution. The number of elements needed in the plan will depend in part on the size of the institution, the species and numbers of animals housed, and the nature of the research being conducted. It is not essential that each institution have a full-time occupational health professional on staff. It is essential, however, that each institution have an established relationship with a professional who can provide the necessary expertise for plan development and operation.

ADMINISTRATIVE PROCEDURES

The complexity of using nonhuman primates in research requires a seamless integration of several institutional positions and programs, including the institutional official, the IACUC, the vivarium management team, the environmental health and safety program, the occupational health and safety program, and the investigator. The close phylogenetic relationship of human and nonhuman primates and the infectious agents endemic in many primate populations require that access to nonhuman-primate colonies be restricted in order to protect both human and animal health. Controlled access depends on administrative support from the institutional official and associated support staff.

There must also be administrative support to enable personnel to conduct training and safety programs. Smaller facilities that do not have the resources to support training and safety staff positions should establish effective systems for employee training and safety monitoring. Occupational health and safety and environmental health and safety consultants should be thoroughly familiar with the occupational hazards associated with working with nonhuman primates, particularly with respect to infectious hazards, such as that represented by B virus.

Finally the institution should make a commitment to the costs associated with tuberculin testing, employee screening, vaccinations, and health assessment for respirator use when required. These costs may be addressed through a process of direct cost recovery or through the allocation of indirect costs from research grants. The critical point is that the elements of the occupational health and safety plan are incorporated into the costs for the animal care program. Some procedures are required by law; others may be elective, but all are prudent for any comprehensive occupational health and safety plan. Table 7-1 provides a checklist for essential administrative features of an occupational health and safety plan.

FACILITY DESIGN AND OPERATION

The design and operation of a nonhuman-primate vivarium are critical features of an occupational health and safety plan. Although well-trained staff and efficient SOPs can address errors in facility design, such errors can have long-term consequences for cost, efficiency, and, in the most serious outcome, risk to employees. The facility design issues for nonhuman-primate vivariums are, in general, similar to those for more generic facilities, although some aspects require increased attention when larger nonhuman primates are housed. Security is paramount. Controlled access through conventional locks, key cards, or other devices is critical. Entry by personnel who have not completed occupational health screening can present a risk to the animal population, but, more important, can present an immediate risk of physical injury to people unfamiliar with the strength or reflexes of nonhuman primates.

Nonhuman-primate species require varying cage sizes and complexity of caging systems. Many of the species used in research require large heavy cages, which can present substantial ergonomic hazards for employees; facility design and caging systems should incorporate features that minimize these hazards. For example, the use of rolling racks and cage elevators for wall-hung cages can help to reduce muscle and back strain. If the husbandry system requires animal transport via cages or tunnels, these should be designed to minimize heavy lifting above shoulder level. Hydraulic lift tables on wheels can be very useful for large primates such as baboons and chimpanzees. These can ease transport of large primates, sometimes dropping to within 4 inches of the floor.

Wall and floor surfaces should be constructed of materials that are resistant to chemicals and cleaning agents and may be easily cleaned.

Floor surfaces should be designed to provide traction to help personnel avoid injuries from slips and falls. Special attention should be paid to floor surfaces in such support facilities as procedure rooms, the cagewashing facility, the veterinary clinic, and the necropsy area.

Sinks for hand washing, eye wash stations, and showers should be placed to ensure ready access for employees involved in chemical splashes or spills. Employees should also have ready access to disinfectant stations, bite-scratch kits, and emergency kits.

Locker rooms, break areas, and employee lounges should be designed to minimize cross-contamination between these areas and the employees' personal clothes and food items. The separation of clean areas from "dirty" areas of a locker room by some sort of stepover design is optimal. All personnel working in nonhuman-primate facilities should have access to a shower. Lounges should be readily accessible to ensure that employees do not eat, drink, or smoke in animal areas.

The size and complexity of nonhuman-primate housing areas require provisions for power failures or other mechanical breakdowns. In addition to basic considerations of animal care, these provisions are critical for employee safety. Emergency lighting helps to prevent accidents and severe injuries during blackouts. In quarantine facilities or containment facilities (for example, ABSL 3 holding rooms), negative air pressure is essential to prevent worker exposure to pathogens or toxic compounds. Such facilities should have redundant power and mechanical systems. Table 7-2 provides a checklist for facility design and operations.

EXPOSURE-CONTROL METHODS

Exposure control is key to the safe operation of nonhuman-primate vivariums. Exposure control is developed in a hierarchic structure to ensure worker safety: engineering controls, work practices, and personal protective equipment. Each element is an important part of the safety plan. The ideal is that no potential exposure route is limited by a single control. Rather, engineering controls, work practices, and personal protective equipment should provide a layered safety net to prevent worker injury.

Engineering Controls

Engineering controls are integrally related to facility design and operation such as directional air flow and double door access barriers. Engineering control features for animal rooms and laboratories also include biosafety cabinets to limit aerosol exposures, chemical fume hoods in laboratories to limit exposure to chemicals, covers on electrical outlets where water is used to wash down rooms, covers on the cage wash pit, and downdraft tables in necropsy suites. Design and operation of the vivarium should minimize repetitive motions and activities that can lead to ergonomic injuries; this is particularly relevant to such activities as cage transport, animal transport, and animal restraint. Table 7-3 provides a checklist for engineering controls.

Work Practices

Optimizing work practices for employee safety is a matter of providing the engineering controls described above and integrating them with employee training and facility SOPs. Strict adherence to safe work practices is a key element of employee safety.

The first element of safe work practices is personal hygiene. Several studies (Gopal and others 2002; King and others 1999) have identified hand washing as essential for preventing infections in human hospitals and animal facilities. Employees should be familiar with routes of infection and recognize that handling objects with gloves and then touching persons or objects with the outside of the gloves creates the potential for personnel exposure to pathogens. Personal hygiene should be encouraged by providing workers with dedicated clothing or protective wear, such as jump suits, laboratory coats, and other clothing that can be left at the workplace.

Housekeeping is also a basic element of safe work practices. A clean, uncluttered work area facilitates sanitation and disinfection and minimizes chances of personnel exposure to pathogens. Keeping animal rooms and laboratories free of clutter reduces the potential for falls and injuries. Some measures, such as sticky mats and footbaths, may be useful in specific circumstances but must be accompanied by a commitment to maintenance and regular replacement.

Sound work practices can optimize worker safety in the cleaning of cages. Use of low-velocity hoses to minimize aerosol formation can help to reduce potential exposures but must be balanced with husbandry needs. Dry cleaning and cage-pan removal should be evaluated by the safety officer and management to minimize ergonomic stress and optimize infection control.

SOPs and training are essential in the handling and transport of animals. Transport of animals depends on appropriate caging or transport boxes, which must be designed to house the animals properly while minimizing ergonomic and infection hazards. Specialized restraint equipment should be used only if the workers and the animals have been trained. Animals should be handled directly only when under anesthesia, if this is possible. If animals are to be captured in a net or restrained by hand, personnel should be provided with appropriate safety equipment and training. If animals are to be transported through common use hallways and facilities (e.g. transport to an imaging facility in a hospital), SOPs should be developed for how to safely transport the animal, as well as a response to exposure plan for non-animal using individuals that are exposed during transport. SOPs should also be developed for cleaning of common use equipment (e.g., MRI machine) and facilities.

The research environment often includes sharp instruments, such as needles, scalpel blades, and catheters. Needle-less or protected-needle systems should be used whenever possible to reduce the potential for injury and exposure to pathogens. Safe work practices should include appropriate signage, and provision of containers for sharps disposal.

Waste from nonhuman primates should be objectively assessed by the safety committee to ensure proper disposal of infectious material generated in either routine husbandry or research.

Employee training should include safe operation of steam autoclaves and other equipment designed to decontaminate infectious waste generated in the vivarium as well as appropriate use of biohazard bags for disposal. It should also include appropriate disposal procedures for radioactive waste and hazardous chemicals. Table 7-4 provides a checklist for safe work practices.

Personal Protective Equipment

The final element of worker safety is the proper use of PPE. Many nonhuman primates are intelligent and very quick—they can intentionally or unintentionally inflict severe injury necessitating the need for engineering controls and SOPs even in the absence of potentially infectious hazards. Worker safety is dependent first upon facility design and appropriate caging systems, but PPE is also essential. It should not be viewed as the sole element or used as a substitute for proper facility design, appropriate equipment, and safe work practices. Determination of appropriate PPE will depend on the nonhuman-primate species, the work environment, and the type of research being conducted.

The minimal personal protective equipment for working with nonhuman primates should be dedicated clothing, gloves, and mask. Workers must be trained in the proper use of these and other personal protective equipment.

Eye and mucous membrane protection are critical in nonhuman-primate work settings. Various kinds of protective eyewear and face shields are commonly used in many laboratories and vivariums, including nonhuman-primate vivariums. The function of this personal protective equipment is primarily to prevent exposure from droplets, projectiles, and chemicals in the workplace.

As early as 1988, recommendations were issued by the CDC that "masks and protective eyewear or face shields" should be used by personnel working with nonhuman primates either naturally or experimentally infected with simian immunodeficiency virus (CDC 1988). In the NRC report *Occupational Health and Safety in the Care and Use of Research Animals* (NRC 1997), it is recommended that "personnel who work with nonhuman primates should wear face shields and other protective garments and equipment appropriate for the circumstances and species involved." Following a fatal human case of B virus encephalitis caused by an ocular exposure to body fluid from a rhesus macaque, the NIOSH and the CDC issued recommendations that protective eyewear should be mandatory for individuals working with macaques (CDC 1998; NIOSH 1999). The CDC issued further recommendations on eye and face protection in the CDC-NIH document *Biosafety in Microbiological and Biomedical Laboratories* (CDC-NIH 1999), where it is recommended that appropriate eye and face protection be based on risk assessment in the setting at hand.

In view of these differences between various published recommendations and the differences in usual practice in various nonhuman-primate facilities, it seems prudent to define a common practice standard, a common recommendation suitable for universal use. This has been done, as follows:

- Eye and face protection should be mandatory for individuals working with macaques.
- Eye and face protection are highly recommended for individuals working with other Old World monkeys and apes based on a splash exposure assessment, on the recognition of human infection by other viruses such as SIV.
- Eye and face protection for individuals working with other nonhuman primates should be decided on the basis of sound institutional review of the hazards, that is sound risk assessment, and appropriate overall risk management protocols and proactive management practices.

The mandatory requirement for eye protection when working with macaque species should be implemented in the context of the institutional animal management program and experimental use. The variety of different housing and experimental environments precludes listing appropriate eye protection for all the situations that may be encountered in a research facility. The use of eye protection with macaques should still be based on an institutional risk assessment and determination of the degree of risk of splashes and mucous membrane or ocular exposure.

When animal husbandry or experimental protocols require working in close proximity with awake primates (manual restraint or restraint equipment) it is appropriate to be conservative in requiring routine use of eye protection.

It is also recognized that new employees or employees with less than 2 years of experience may be at higher risk for injury or exposure (bin Zakaria and others 1996). The higher risk for new personnel may warrant a standard requirement for use of PPE, regardless of tasks being performed. This decision is best made at the institutional level by the process of risk assessment and proactive management.

Importantly, personal protective equipment must protect workers from potential exposures while not compromising their dexterity or vision. Excessive personal protective equipment can present ergonomic hazards and hazards associated with heat stress in work environments that are not temperature-controlled.

All workers must be trained in the use of PPE. If there are specific requirements for respiratory protection, the program must involve an occupational health professional to determine whether the use of a respirator is contraindicated in workers with pre-existing medical conditions or health concerns. The proper respiratory device should be selected and fit testing should also be done, if appropriate (e.g., for N-95 mask) (29 CFR 1910.134). Table 7-5 provides a checklist for PPE use.

EDUCATION AND TRAINING

Worker education and training constitute a core element of an occupational health and safety program. Because this aspect of the program is critical in nonhuman-primate facilities, it is addressed in detail in Chapter 8. Table 7-6 provides a checklist for education and training.

OCCUPATIONAL HEALTH

The occupational health aspect of the occupational health and safety program can be divided into two elements. The first, preventive medicine, includes preplacement medical evaluations, follow-up periodic and episodic health evaluation, appropriate immunizations and serum banking. The second is an appropriate response system in the event of an employee injury. Appropriate first aid and medical care must be immediately available to deal with worker injuries and exposures in the nonhuman-primate facility.

Preventive Medicine

The preventive preplacement medical evaluation typically used to screen workers entering a nonhuman-primate facility originated with the intent to protect the health of the employees as well as the animals (Muchmore 1975). This evaluation may consist of a questionnaire followed by a physical examination if a need is established. Health questionnaires should be confidential and evaluated by appropriate occupational health professionals. Information collected in a questionnaire may include previous or ongoing medical problems, current medications, allergy history, prior immunizations, and previous results of tuberculin skin testing. The evaluation may identify pre-existing conditions that might modify an employee's risk profile (NRC 1997), such as tuberculosis or potential pregnancy in women of child-bearing years. The preplacement medical evaluation may also serve as an opportunity to educate the employee about potential hazards of working with nonhuman primates (NRC 1997). Such evaluations also establish a link for the employee with the appropriate occupational health professional. It is desirable that employees recognize the occupational health professional as a resource in addressing their concerns in the workplace environment.

The recognition of immunocompromised or pregnant persons may be difficult because of patient confidentiality laws and regulations, but such persons are at special risk so all workers must be informed of this (Rayburn 1990). Employees should be advised to communicate with the institutional occupational health professional, who can then communicate with the employee's personal physician. Anesthetic gases, radiation, and certain infectious diseases are well-recognized risks for pregnant employees, and in most cases alternative work assignments are appropriate. Table 7-7 provides a checklist for occupational health issues.

A health screen is also necessary if workers will be required, as part of their daily duties, to wear respirators rated by the National Institute for Occupational Safety and Health. A respirator may rarely pose an additional hazard if workers suffer from heart disease, respiratory illness, or diabetes (Szeinuk and others 2000).

The facility may use periodic health evaluations to evaluate the success of its occupational health and safety program in reducing occupational illness and injury (NRC 1997). The nature and frequency of these periodic evaluations should be based upon the nature of potential hazards; in NHP facilities, these periodic evaluations should be focused on physical injuries as well as illnesses arising from exposure to relevant infectious agents. Mild symptoms of health alterations may be indicative of a need for better preventive measures. The need for and design of periodic health evaluations should be determined by representatives of various oversight or advisory bodies associated with the institution—environmental health and safety program, occupational health service, office of human resources, animal facility director, etc. Decisions regarding the nature and frequency of periodic evaluations should be reviewed regularly, based upon changes in working conditions and exposures, injury and illness experience, and the availability of new guidelines for good occupational health practice (NRC 1997).

Health evaluations should also be conducted in response to persistent symptoms, symptoms that indicate the onset of a work-related illness, or the occurrence of a work-related injury. These episodic evaluations should typically include a physical examination focused on the major complaint, and the employee may need to be referred to medical specialists if the illness/injury warrants it. Work-related injuries/illnesses that lead to medical evaluation and loss of work time should be reported to the occupational health information system (BLS 1986; NRC 1997).

Another important component of the occupational health care system is the immunization program. The decision to immunize an employee is influenced by the potential risks the employee may face on the job and should be determined at the time of the preplacement evaluation, or at periodic or episodic evaluations. As required by the Occupational Safety and Health Administration Blood-borne Pathogens Standard (OSHA 1991), vaccinations must be offered to personnel who will be working with experimental pathogenic agents such as hepatitis B virus. Additional vaccines may be offered for tetanus, measles, and other etiologic agents that are applicable to the research program.

The emphasis on serum-banking as a routine preventive medicine function has decreased over the last decade. A survey of 50 institutions conducting animal research demonstrated limited utilization of reference serum banks. (Lehner and others 1994). Sera from only 6% of personnel were used for epidemiological studies. Only 0.3% of sera stored was used for medical/legal or diagnostic purposes. Serum banking has value only when its purpose is to obtain data for the conduct of occupational risk assessments. Each institution should develop its own plan and assess the utility of an annual serum bank. The plan needs to consider a combination of factors such as chain of custody, confidentiality, resource requirements for long-term storage, and accessibility for serologic testing (NRC 1997). Most institutions do collect serum samples following an employee exposure, such as a bite, needle stick, or scratch where there is potential for occupationally acquired disease.

TUBERCULOSIS TESTING

As an important ancillary function provided by the occupational health service, periodic tuberculosis skin testing using purified protein derivative (PPD) can help identify individuals that could potentially transmit tuberculosis to nonhuman primates and provide important baseline clinical information for future medical management decisions following a workplace exposure. Expert consensus panel recommendations for targeted tuberculin skin testing are outlined in various clinical guidelines (www.cdc.gov/mmwr/; Vol. 49, No. 6) and www.cdc.gov/nchstp/tb/pubs/1376.pdf). It is reasonable to assess all nonhuman-primate handlers upon initial entry into the workplace. A "two-step" tuberculosis skin test is recommended for those employees not previously under periodic PPD testing (Sherman and Shimoda 2001). The frequency of periodic PPD skin testing for nonhuman-primate handlers will depend upon risk assessment for the particular facility. The principle function of employee tuberculosis screening is to protect the nonhuman-primate colony and this should be a prime consideration in determining the frequency of periodic PPD skin testing for employees. Additional considerations should include factors such as the likelihood of tuberculosis infection among the facility nonhuman-primate population, immune status of nonhuman primates being handled, nature of staff contact with nonhuman primates, experimental protocols being employed, and past experience with tuberculosis in the nonhuman-primate facility. An example of one large research institution's recommendations for periodic tuberculosis screening is provided in Box 7-1.

Postexposure Treatment

The proper response to injuries and exposures involving macaques, which present the risk of B virus exposure, is described in Chapter 9. As stated in Chapter 9, this special case, often involving medical personnel (e.g., typical emergency department physicians) with little familiarity with the prevention of human B virus infection, requires adequate preplanning. All institutions housing macaques must have a defined plan to deal with occupational B virus exposures.

EQUIPMENT PERFORMANCE

Several aspects of exposure control depend upon specialized mechanical equipment. Examples are air-handling systems with HEPA filtration, anesthetic machines, biosafety cabinets, fume hoods, cages, cage washers

BOX 7-1
Recommendations for Tuberculin Testing Procedures at the National Institutes of Health (NIH 2001)

1. If the new worker has a previous positive intradermal skin test (with purified protein derivative, PPD), then no further skin testing should be done. Instead, a careful history for symptoms of active tuberculosis should be taken. A chest x-ray should be obtained if the history is suggestive, or if the new worker cannot provide documentation of a normal chest x-ray following the positive skin test, or if the new worker has received inadequate treatment following the positive skin test. If the chest x-ray is negative the worker should be cleared for work. Otherwise further investigation and possible therapy is called for. Until this is completed, the worker should not be allowed to work with nonhuman primates.

2. The new worker without a history of a previous positive skin test should be given a preplacement tuberculin test. A second test, two weeks after the first, is recommended if the first test is negative. If the first skin test is positive, then a careful history for symptoms of tuberculosis and a chest x-ray should be obtained. If the employee does not have a documented negative skin test in the previous 24 months and there is neither clinical nor radiographic evidence of tuberculosis, the worker should be cleared for work but urged to return for periodic follow up. If there is a documented negative skin test in the preceding 24 months and the chest x-ray is negative, the worker should be started on prophylactic treatment and allowed to start work after 3 days of therapy. However, if the chest x-ray is positive then the worker must be evaluated and treated for active tuberculosis. Such an individual should not be allowed to work with nonhuman primates until it is clear that he/she is not contagious.

3. If the first skin test is negative and the second is positive it is indicative of new or recrudescent infection. A medical history for symptoms of tuberculosis and a chest x-ray should be obtained. If both are negative the worker should be urged to return for periodic follow up and should be cleared for work. However, if the chest x-ray suggests active disease then the employee should be referred for further diagnosis and treatment. If both the initial and secondary intradermal skin tests are negative, the worker should be cleared for duty.

4. Periodic skin testing should be done on workers exposed to pulmonary or lymphatic tissues of nonhuman primates. Workers exposed to other nonhuman-primate tissues may be tested, according to a risk assessment.

and autoclaves. Such equipment should be regularly maintained and, if necessary, recertified. Operational logs (for example, recording time/temperature) should be maintained for medical-waste decontamination equipment. Biosafety cabinets and fume hoods are typically subject to regular maintenance and recertification. Personnel should be trained so they can interpret abnormal readings of pressure gauges and other indicators used in containment facilities. Airflow monitors should be used regularly to verify air exchanges in animal rooms and to determine pressure differentials. Cages and equipment for restraining animals must be in good repair and proper functional state. Defective equipment must be taken out of service until repaired. Table 7-8 provides a checklist for equipment performance.

INFORMATION MANAGEMENT

Information management is essential in the development and maintenance of an OHS plan. A written plan is needed to identify the appropriate levels of training for all new employees. Training necessarily includes providing SOPs and written guidelines to new employees. There should be documentation that employees have read these materials as part of their training program. The institution must maintain its injury- and illness-prevention plan in accordance with state and federal regulations. Training must be documented to provide regulatory agencies and other oversight bodies with evidence that appropriate training is carried out regularly.

Annual retraining is required in some circumstances by law or regulation. The retraining venue helps ensure that workers are familiar with changes or additions to the OHS plan and SOPs. The retraining venue is also an appropriate venue for reminding workers to review changes in their own health status; for example, a change in the condition of one's respiratory system or the presence of a new immunosuppressive disorder may necessitate a change in duties or a review by the occupational health physician.

The institutional injury log is a critical database in assessing the efficacy of the occupational health and safety plan. Analysis of injury rates for different tasks and different types of housing systems can be invaluable in deciding whether to modify the OHS plan. The institutional injury database should include data that allow matching of risks to job classifications. By including data on both injury rates and populations at risk, occupational health and safety program professionals are able to determine the efficacy of their programs. The institutional injury database and safety inspection program must be integrated in a system that favors full follow-up and correction of problems. If identifying information is included in the injury database, this database should remain confidential and accessible only by the appropriate institutional officials. Table 7-9 provides a checklist for information management.

EMERGENCY PROCEDURES

Every nonhuman-primate facility should have an emergency action (disaster) plan. The plan must include the means for dealing with fires, power failures, earthquakes, floods, tornadoes, and other life-threatening emergencies. A specific plan should be developed for procedures to be used in the event of an animal escape; this plan should minimize employee hazards and risks. The emergency action plan should identify the responsibilities of in-house personnel and provide readily available lists of telephone numbers of additional expertise and resources. Practice drills of elements of the emergency action plan are necessary; this is especially the case for an animal escape. The proper response to injuries and exposures involving macaques, which present the risk of B virus exposure, is described in Chapter 9. As stated in Chapter 9, this special case, often involving medical personnel (e.g., typical emergency department physicians) with little familiarity with the prevention of human B virus infection, requires adequate preplanning. All institutions housing macaques must have a defined plan to deal with occupational B virus exposures. Table 7-10 provides a checklist for emergency procedures.

PROGRAM EVALUATION

The efficacy of the OHS program should be evaluated regularly. At a minimum, the IACUC and occupational health professionals should review the OHS plan as part of their semiannual reviews. Review of injury and illness logs should be a routine responsibility of the occupational health and safety program officer or a designee of the institutional official (for example, the director of the animal program) in the case of smaller facilities. If the nonhuman-primate facility is part of a large institution, it is often valuable to involve the institutional occupational health and safety program staff in this review. This provides a new perspective on program design and implementation and complements the perspective of the facility manager and senior staff.

Effective program evaluation requires appropriate expertise to address risks associated with hazards that cannot be eliminated. It also requires full involvement of supervisors at all levels. This kind of program evaluation helps to develop a culture of safety throughout an institution. Table 7-11 provides a checklist for program evaluation.

TABLE 7-1 Checklist for Essential Administrative Features of the Occupational Health and Safety Program of a Nonhuman-Primate Facility

	Program Elements
	Designated official responsible for the program.
	Clearly defined organizational structure with assigned responsibilities.
	Integration of safety issues dealing with nonhuman-primate biology in Institutional Animal Care and Use Committee protocol and program review.
	Institutional or consultant expertise regarding hazards associated with nonhuman primates.
	Personnel training regarding nonhuman-primate behavior, hazards, and risk reduction.
	Established lines of communication among administrative components of the research program (Institutional Animal Care and Use Committee, Environmental Health and Safety, Occupational Health and Safety Program).
	Designated safety officer.
	Institutional occupational health professional.[a]
	Semiannual IACUC review and report to institutional official.

[a]Some institutions allow staff to use a personal physician. If this is the case, the OHP should require a feedback loop from the personal physician to occupational health staff to ensure tracking of occupational illnesses or injuries as well as possible modification of job duties based upon that feedback.

TABLE 7-2 Checklist for Facility Design and Operation

	Program Elements
	Facility design minimizes the presence of personnel in nonhuman-primate areas and reduces potential for accidental contact between visitors and nonhuman primates.
	Restricted access to nonhuman-primate housing and the presence of viewing ports on entry doors or anterooms to allow observation of animal room before employee entry (to check for escaped animals).
	Squeeze back cages or other equipment to facilitate manual restraint.
	Cage design to minimize ergonomic hazards in routine husbandry.
	Adequate room size to allow personnel space to work with one cage and maintain adequate distance from other cages.
	Adequate air exchange and appropriate air balance for containment and hygiene.
	Appropriate construction materials in animal areas to resist impact of heavy equipment and caustic chemicals.
	Animal room doors designed to be opened from the inside without a key.
	Appropriate locker facilities and showers for personnel.
	Adequate break areas for employees.
	Laboratory facilities should be at the appropriate ABSL level, equivalent to biosafety levels in the vivarium. For example, if there are ABSL-3 animal studies, are the clinical laboratories equipped for ABSL-3 samples?
	Sinks readily accessible in vivarium for hand washing, readily accessible locations for PPE and readily accessible eyewash stations and emergency and bite-scratch kits.
	Dedicated quarantine facility with anterooms, negative air pressure and autoclave for Centers for Disease Control and Prevention quarantine program.

TABLE 7-3 Checklist for Engineering Controls

	Program Elements
	Properly certified biosafety cabinets must be available for tissue processing and procedures that may create aerosols.
	Routine maintenance schedule for ventilation system.
	Splash guards for personnel operating laboratory equipment such as hematology analyzer and centrifuges.
	Covers on electrical outlets where water is used.
	Downdraft necropsy tables to reduce aerosol exposure.
	Floor materials designed for durability and safety during the sanitation procedures required in nonhuman-primate vivariums.

TABLE 7-4 Checklist for Safe Work Practices

	Program Elements
	Hazardous areas are clearly posted with signage.
	Nonhuman-primate areas have access restricted to essential personnel.
	SOPs are in place to identify essential personnel, and integrate appropriate safety practices in routine work practices.
	SOPs in place for infection control and decontamination of work surfaces.
	SOPs in place for personnel hygiene, including eating, drinking, smoking, applying cosmetics, appropriate work clothing, and personal protective equipment.
	SOPs in place for appropriate handling and disposal of hazardous materials (e.g., urine, feces, carcasses, animal tissues, veterinary supplies).
	SOPs in place for transport of animals within the facility and into and out of the facility.
	SOPs in place for safe restraint of nonhuman primates.
	Adequate personnel available for tasks requiring restraint and handling of awake nonhuman primates.
	SOPs in place for cage cleaning to minimize potential hazards to personnel.
	Needle-less technology used when possible. Proper receptacles used for disposal of sharps.
	SOPs in place for appropriate personal protective equipment for specific tasks.
	Primate housing should be consistent with biosafety standards specified in *Biosafety in Microbiological and Biomedical Laboratories* (CDC-NIH 1999).
	SOPs in place for work practices that are appropriate to the nonhuman-primate species being used.
	SOPs in place for behavioral training of nonhuman primates for routine procedures and experimental manipulations.
	SOPs for wound management.

TABLE 7-5 Checklist for Personal Protective Equipment

	Program Elements
	Selection of appropriate personal protective equipment should be based on hazard identification, assessment of risks associated with specific tasks, and the level of training of individual workers.
	Personal protective equipment should be provided by the institution, readily available, and well maintained.
	Personal protective equipment should be readily available for maintenance and service personnel who must access animal holding areas. These people should be trained in personal protective equipment use.
	Selection of specific personal protective equipment should be made in keeping with the nonhuman-primate species involved and known zoonotic infectious agents carried by that species.
	SOPs for personal protective equipment should include specific considerations for hazards present in the work environment. These may include tasks involving use of sharp instruments, creation of aerosols, and activities that may result in close proximity to behaving primates.
	Personal protective equipment should protect the mucous membranes of the eyes and mouth from contact with splashes and sprays of contaminated fluids and chemicals.
	Institutional policies for personal protective equipment should include consideration of personnel working in situations without environmental controls. Personal protective equipment requirements should minimize potential risks of overheating and obscured vision.
	Occupational health and safety professional and/or training officer available for proper training and certification of personnel in use of personal protective equipment.
	Occupational health and safety professional monitors and audits personal protective equipment use and disposal.

TABLE 7-6 Checklist for Education and Training[a]

	Program Elements
	All employees or visitors (including students) who enter nonhuman-primate areas, or handle nonhuman primates or their tissues must be included in the training program and in the occupational health and safety program. Completion of training must be documented. It is the responsibility of the institutional official to assure that there is conformance with this requirement.
	All employees or visitors (including students) who enter nonhuman-primate areas, or handle nonhuman primates or their tissues must be informed of all infectious and environmental hazards, whether experimentally introduced or generally associated with the nonhuman primates housed in the facility.
	Employees must have ready access to SOP manuals and all appropriate material safety data sheets and biohazard information.
	The Institutional Animal Care and Use Committee should maintain records of personnel qualifications and training or ensure that adequate institutional records are kept. This could be the manager but ultimately it is the responsibility of the IACUC to ensure this is in place.
	Husbandry and research staff should be trained on appropriate responses to emergency situations. This includes review of procedures for fires, chemical spills, and animal escapes.

[a]Greater detail of the education and training component of an OHP follows in Chapter 8.

TABLE 7-7 Checklist for Occupational Health Issues

	Program Elements
	A program of routine tuberculin skin testing in place for all personnel.
	If routine serum banking is incorporated into the program, a system for proper storage and chain of custody for the samples should be in place.
	Policies, developed in collaboration with the occupational health professional, should be in place to control access to the NHP facility by immunocompromised employees, pregnant women, and children. The policies should generally prohibit access by children under 18 years of age.
	Appropriate blood tests and/or vaccination programs should be available to employees as determined by the occupational health professional or as required by regulatory agencies.
	Polices for inclusion of all personnel (e.g., students, visiting scientists, nonaffiliated members of IACUC, physical plant staff) in the occupational health and safety program and for preplacement examinations or medical record reviews should be in place.
	Where N-95 respirators are used there must be an appropriate health clearance system and training and fit testing of equipment.
	Programs should provide occupational health professionals with a mechanism for counseling employees. This should allow employees to address questions or concerns in a confidential environment.
	Occupational health staff should be familiar with facility design, routine procedures, and infectious and chemical agents in use at the facility.
	Medical personnel providing occupational health support must be trained and knowledgeable regarding the risk and response necessary to deal with B virus exposure.

TABLE 7-8 Checklist for Equipment Performance

	Program Element
	Annual certification programs for equipment including biosafety cabinets, anesthetic machines, autoclaves, and fume hoods should be in place (or per manufacturer's recommendations).
	There should be routine maintenance and repair for all air filtration (HEPA filters) and air handling equipment and backup generators. Redundant systems may be necessary in some areas.
	There should be routine maintenance and repair of emergency showers and eyewash stations.
	Logbooks and routine testing should be in place for any equipment used in decontamination of biomedical waste.
	Animal restraint equipment should be easy to operate and on periodic maintenance schedules to reduce ergonomic hazards and employee injuries.
	Cages and restraint equipment should be inspected regularly and resources available to repair defective equipment.
	Sewage entering the liquid waste stream must comply with all regional environmental regulations.

TABLE 7-9 Checklist for Information Management

	Program Elements
	A program of employee safety training should be in place; this should be documented, including a record of course content.
	Employee training documents should include information identifying specific hazards, risks of injury from those hazards, and appropriate measures to reduce those risks.
	There should be an injury / illness log that is reviewed on a periodic basis to provide feedback on efficacy of the overall safety program.
	Efficacy of employee training programs should be assessed by formal mechanisms of testing and safety audits conducted by the safety officer.
	Results of inspections should be documented, including corrections of hazardous conditions or practices.

TABLE 7-10 Checklist for Emergency Procedures

	Program Elements
	Information about the appropriate response to animal-related injuries should be available to all employees at any time.
	SOPs should be in place for sample collection and analysis in the event of personnel injuries or exposures involving B virus.
	Bite-scratch and emergency kits, if retained by an institution, should be properly and conveniently placed in all areas where nonhuman primates or their associated equipment are encountered. These kits should be inspected on a regular basis to ensure the contents are not beyond the expiration dates.
	There should be routine maintenance for emergency showers and eyewash stations.
	There should be a standard information format for collection of information related to nonhuman-primate emergencies and injuries.
	Medical care should be available for employee-related injuries at any time, including weekends and holidays.
	A disaster plan should be written, distributed, and readily available.
	Local emergency preparedness agencies should be aware of facility disaster plans and biohazards.

TABLE 7-11 Checklist for Program Evaluation

	Program Elements
	Injury and illness statistics should be reviewed on a periodic basis to determine any areas of program deficiencies and opportunities for improvement.
	The safety committee should meet on a regular basis and report results of program evaluation and deficiencies to the institutional official.
	Program deficiencies that affect employee safety should be addressed in a timely manner by the institutional official.
	Facilities that are part of larger institutions should be subject to safety audits by host institutional safety programs.

8

Personnel Qualifications, Training, and Continuing Education

INTRODUCTION

Research, educational, and testing facilities that use nonhuman primates are usually dynamic environments. Changes in these environments require a commitment to training and continuing education to provide a safe and healthful workplace. Some changes involve staff turnover and the involvement of new students and visitors at a facility. New animals and even new nonhuman-primate species are received and new studies begin with their own procedures and equipment. Other changes are facility-related, such as those caused by facility modification (addition or reconfiguration). Changes in equipment can include new caging and housing systems, new materials-handling equipment, new anesthetic or surgical equipment, and the use of new instruments and devices. Training and continuing education programs provide an opportunity for management and staff to communicate about workplace hazards, program changes, updated organizational policies and procedures, and emergency procedures.

An institution can achieve its OHS objectives only if its employees know the hazards associated with their work; understand how the hazards are controlled through institutional policies, engineering controls, work practices, and PPE; and have sufficient skills to do their work safely and proficiently (NRC 1997). To meet those requirements, a multifaceted training program that addresses the full array of health and safety issues related to the care and use of nonhuman primates must be instituted.

Such a program not only benefits the workers, it also benefits the institution through minimization of occupational illness and injury and lost time.

Institutions have legal and ethical responsibilities to provide staff with the training, knowledge, and equipment they need to protect themselves from workplace hazards. In 1985, Congress enacted two laws that require that institutions provide training for staff who care for or use animals: the Health Research Extension Act (Public Law 99-158), which made compliance with the Public Health Service (PHS) Policy a matter of law for all PHS-funded research, and the Food Security Act (Public Law 99-198), which amended the Animal Welfare Act (7 USC 2131-2156).

This chapter provides a framework for assessing personnel qualifications and developing training and education programs for organizations that use nonhuman primates in research, education, and testing.

PERSONNEL QUALIFICATIONS

AAALAC International requires accredited institutions to describe personnel qualifications in the accreditation application; OLAW lists evaluation of personnel qualifications as an IACUC oversight responsibility.

The point of responsibility for assurance of personnel qualifications must be clear if conformance with AAALAC International and federal guidelines are to be met. Nonhuman-primate housing facilities are often centralized and used by research staff from an entire institution. If personnel that work with the nonhuman primates do not share local lines of authority, it can lead to confusion about responsibility for qualification assurance. The facility director is ultimately responsible for determining the qualifications of facility employees, contract staff, support-services staff, program inspectors, and visitors who work with or around nonhuman primates and the associated equipment. Principal investigators are responsible for the verification of the qualifications of research assistants, collaborators, and guests.

Delineation of responsibility for assurance of qualifications should be established by institutional policy. Such policy demonstrates that upper management recognizes the importance of personnel qualifications, has determined an appropriate assurance plan, and has delineated responsibility for the implementation of the plan.

Management should determine operationally specific minimal qualifications for all staff and visitors that work directly with the nonhuman primates, their byproducts, housing, holding rooms, equipment, or tissues. Minimal assurance of qualification is important not only for new employees but also for current staff as they gain proficiency and engage in more challenging tasks. Attention should also be given to the development of qualification standards for support staff, such as maintenance personnel, housekeeping staff, materials-handling staff, providers of such services as pest management, laundry, environmental monitoring, and occupational medicine, and emergency personnel such as police and firefighters. Conducting a facility orientation and an overview of the operationally specific hazards, risks, policies, and emergency procedures for support staff is recommended.

Managers, principal investigators, and others often view possession of pertinent professional credentials, a history of work with nonhuman primates, or simply length of employment as evidence of qualifications. This can be a dangerous assumption. All personnel who work in nonhuman-primate facilities and everyone who works with nonhuman-primate tissues must have knowledge of task-specific hazards, be able to recognize when an exposure has occurred, and be able to accurately demonstrate emergency steps to take in the event of an exposure.

An assessment of basic skills is often a good place to start in assessing a person's qualifications. Through observation, supervisors can note whether an employee has the basic communication skills necessary to perform the task at hand and whether the employee's reading skills are sufficient for understanding the SOP, organizational policies, and material-safety data sheets applicable to their position. After establishing a person's basic skills, management can strengthen assurance of qualifications by conducting behavioral observations and comparing them with the previously established minimal qualifications. Supervisors can observe new and current employees engaged in tasks and note their familiarity with the nonhuman-primate species used, equipment, and tasks. Through behavioral observations, a supervisor will be able to observe an employee demonstrating (or failing to demonstrate) the knowledge, skill, and deportment needed to conduct specific tasks safely.

It is important to assess behavior in a task-specific context. To determine whether a person is qualified to perform a specific task involving a nonhuman primate or nonhuman-primate tissues, the supervisor may choose to seek answers to several pertinent questions, such as the following:

- Has the person performed this task before?
- If so, how long has it been since the person last performed the task? Has anything about the task changed since the person last performed it?
- Was previous performance of the task of a similar duration?
- Has the person worked with the necessary equipment before?
- Is the person familiar with the equipment's uses and associated risks?

- Has the person performed this task with others or independently?
- How confident is the person of his or her ability to conduct the task safely?
- Has the person been observed and found to be proficient in the performance of the task?
- Has the person worked with this species of nonhuman primate before?
- Is the person not only familiar, but also experienced, with the species-specific behaviors, housing, handling, and restraint methods?
- If the animal being manipulated is to be anesthetized, does the person show proficiency in the use and administration of anesthesia?

It is clear that a qualification assessment can benefit enormously both from the exchange of information between management and staff and from observation. Thoughtfully constructed dialogue can help to identify skill or knowledge deficiencies and pinpoint the need for initial and refresher training. Supervisors should always verify the written and stated qualifications of an individual employee with observation. This measure will help to ensure that the knowledge and skills that a person claims to have are commensurate with the specific task requirements and safe work practices.

There are many approaches to determining whether a person is qualified to perform a specific task. One approach is the use of proficiency assessment documentation. The following form is an example of how an institution can document an individual's qualifications and proficiency for a specific task. This sample form is aimed at a task or procedure involving awake nonhuman primates, such as a pole and collar transfer or transfer of the animal to a clean cage. The form can be easily modified to accommodate additional tasks, such as cagewashing, material handling, environmental enrichment, and tasks involving anesthetized nonhuman primates.

TRAINING

The ultimate goal of safety training is to reduce occupational exposures, accidents, near-accident incidents, injuries, and illnesses (Gershon and Zirkin 1995), that is, to ensure that safety-related human performance and training discrepancies are identified and that appropriate techniques are used to improve the safety and health of the workforce. The program will do this by providing information to improve knowledge, demonstrating safe work techniques, providing instruction on emergency response, providing information on regulatory controls, providing infor-

Qualifications and Proficiency Assessment Record

Name:________________________ **Supervisor:** ____________________

Date of initiation: ____________________________

State the task involving a nonhuman primate: _________________________
____

State the steps to the task (reference SOP#):____________________________
____
____
____
____
____

Species of nonhuman primate:____

Risk Type: (circle all that apply)

Bite Scratch Splash Percutaneous Other (specify)

Prescribed Personal Protective Equipment and Clothing: (circle all that apply)

Head Cover	Safety Glasses	Goggles	Face Shield
Respirator	Mask	Latex Gloves	Neoprene Gloves
Leather Gloves	Gauntlets	Laboratory Coat	Jumpsuit
Long Sleeve Scrubs	Disposable Sleeve Protectors	Shoe Covers	Steel-tipped Boots or Other Designated Shoes

Other____

Equipment used: (circle)

Chair Manual Restraint Pole and Collar

Transfer Box Other____

Training Progress*:

Date	Duration	Trainer	Comments	Supervisor's Initials

**Number of training sessions and length of time will vary depending on baseline skill and complexity of the procedure. Take the time necessary for the individual to acquire task proficiency.*

Certification of Proficiency:

I, ____
(supervisor)

hereby certify that ____
(trainee)

has demonstrated proficiency in the above named task/procedure.

Supervisor signature:____

Date:____

mation and yearly updates on safety policies and procedures, and motivating staff to work safely (Gershon and Zirkin 1995).

Requirements established by federal and state regulations, accrediting, and funding agencies place various training demands on institutions where nonhuman primates are housed and used, and training records should be maintained and retained in accordance with applicable federal and state regulations and guidelines. Thoughtfully constructed and conducted staff training can have a major impact on the safety record of the institution.

An effective training program requires resources, administrative support, recordkeeping, and a system for monitoring training efficacy. Management can demonstrate its commitment to its safety and health training program by funding it appropriately and by visibly supporting the program, its goals, and its objectives.

Safety training should begin at each new employee's orientation and progress continuously as more complex tasks are assigned and job responsibilities increase. Staff may perform a variety of tasks, from routine daily care and feeding of animals to animal handling and research-associated manipulations; the multifunctional aspect of the work can mean that the skills personnel must have will vary greatly in regard to complexity. Periodic refresher training is needed to maintain proficiency and adherence to institutional procedural standards. Safety training should also be conducted when unsafe or risky behavior is observed.

To be successful, safety training should be delivered routinely. When safety training is provided only episodically, such as when an accident has occurred, it can be interpreted by staff as punitive and have an undermining effect on accident-reporting compliance. Safety training should be conducted during regular work hours; this demonstrates that management values the training and recognizes it as an integral part of employment. Training topics should be chosen carefully to meet real training needs. Training should be developed for individuals and for specific worker groups, and that training should be delivered in terms related to the specific environment and at a trainee-appropriate cognitive level. If possible, trainers should be selected from current staff to ensure the program specificity of the training and to facilitate follow-up discussion after the training. Two-way communication is important in a training program; staff must be afforded the opportunity to ask questions and clarify their understanding of the messages being presented.

Several key factors can improve the effectiveness of training:

- The trainer must be knowledgeable and an effective communicator.

- The message to be conveyed should be succinct and not cover too many topics at once.
- The audience must comprehend the message; training should be aimed at the proper level of knowledge and understanding (including language).
- The environment in which training is conducted should be conducive to the lesson at hand, lighting, temperature, no distractions, and so on (Gershon and Zirkin 1995).

Many OHS professionals have found value in an annual training plan. Upper management, organizational safety committees, supervisors, workers, and safety managers should provide input pertinent to a comprehensive and operationally specific annual training plan. The plan should outline training-specific details, such as training topics, dates and duration of training sessions, needed resources, and training objectives. A training plan communicated well ahead of time to managers and administrators has several benefits; it will minimize organizational disruptions by allowing management to plan around presentations and it provides administrators with information needed for resource allocation and management decisions.

Several methods are available to assist managers and supervisors in their approaches to training. The following elements will help to ensure that the instructional program will be effective and meet the needs of both the organization and individual personnel: needs assessment, development of objectives, determination of content, selection of methods and techniques, establishment of the timing and administration of training, and evaluation are all important (Broadwell 1986). These elements are discussed below.

Periodic Needs Assessment

Training needs should be assessed in regard to organizational needs, worker group needs, and individual worker needs. Organizational needs are based on organizational goals and objectives. Worker group needs can be described as departmental or unit needs, such as the needs of a research group or the needs of those involved with routine husbandry procedures, including care and feeding. Examples would be group skills training or familiarization with required procedures for a new study and training in the use of a new piece of equipment. Individual employee needs can be as varied as basic literacy and numerical skills, special technical skills, new-employee orientation, and task-specific employee competence. In many occupations, the confidence that an employee has in his or her own ability can directly influence on-the-job safety. Managers and supervisors should listen to staff, observe them, and be responsive when employees demonstrate or otherwise convey their lack of confidence in performing a particular task.

Before launching into training as a solution, it is important to determine whether training is likely to solve a problem or resolve an issue; training is not the answer for all situations. There is often a difference between what people are expected to do and what they are actually doing, and training might not be effective in resolving the disparity. For example, a manager might observe an increase in the number of occupational injuries associated with the movement of nonhuman-primate caging. If investigation reveals that employees are not using a mechanical lift to move large objects, such as macaque caging, and thus are not following SOP, management needs to look into why the staff are not using the lift. Talking with staff directly can often reveal why they are performing outside operational standards. In the example, such discussion could include the following questions:

- Is the lift operational?
- Is the lift the right piece of equipment to do the job?
- Are there enough lifts available to meet the unit needs?
- Does corridor activity permit the safe passage of a mechanical lift?
- Do the employees know how to operate the lift correctly?

Such questions will help supervisors and managers to determine whether training is warranted. Perhaps the real reason for not using the lift is simply that at peak use times there are not enough lifts to meet staff needs. Intervention might include the purchase of enough lifts to meet staff needs, rather than lift-operation training.

Work analysis, job and task inventory, assessments, and surveys or questionnaires can be useful for pinpointing training needs. A work analysis is accomplished by breaking down a job into the required skills and knowledge. Similarly, a job or task inventory involves breaking down a job and ranking tasks by whether lack of pertinent knowledge or skill can have serious consequences. Assessments or surveys can be conducted by written or oral methods and can reveal training needs through identification of discrepancies between employee replies and the responses desired by management. Surveys can be conducted anonymously to give a trainer a feel for group beliefs, knowledge, or perceptions of risk, or on an individual basis to provide a trainer with individual specific training needs.

A needs assessment will identify discrepancies between what people are doing, how they are doing it, and the prescribed way to do it safely. A needs assessment should involve an overall evaluation of an organization's program, processes, and past performance. It is most helpful to assemble an interdisciplinary group of people to conduct the assessment or to seek their individual inputs. The functional parts of the overall program should be examined initially followed by the processes or activities that define them. In a research facility that houses nonhuman primates, the functional parts might be animal care and feeding, enrichment, animal health, and technical assistance. Activities associated with animal care and feeding would include cagewashing, animal transfer, material handling (caging, feed, and bedding), and daily animal room maintenance. A review of past performance is one way to evaluate a quality assurance program and is also fundamental to a complete needs assessment: injury and illness reports, absentee and out-sick records, safety and health citations, and staff and management views about worker safety and health concerns. This review can be inserted into the quality assurance program.

Development of Objectives

Training objectives should clearly reflect the desired training outcome. There are programmatic objectives and learning objectives. Both should be identified in the training plan. The programmatic objectives should be written to meet organizational goals, for example, to reduce accidents by 10% on an annual basis. Learning objectives should state what is expected of the trainee. Properly written learning objectives have three basic elements: the task to be completed, the conditions under which the task is to be completed, and the behavior or performance that is being measured and the standard against which it will be measured (Broadwell 1986). These three elements may be expressed as performance, condition, and criteria (Mager 1997). The objectives must be specific, measurable, attainable, realistic, and trackable (Broadwell 1986). The following are examples of learning objectives:

- The trainee will demonstrate how to perform an eyewash at the unit's eyewash station.
- The trainee will be able to quickly list three areas of the face that are recognized as sites of a mucosal exposure to potentially infectious splashes.
- The trainee will be able to quickly state the steps to take after a percutaneous exposure to material potentially contaminated with B virus.
- The trainee will be able to state one aggressive macaque behavior and one nonaggressive macaque behavior as taught in class.

Determination of Content

Training content should reflect primarily what the trainee must know as opposed to what the trainee should know or could know (Broadwell 1986). To enhance comprehension and retention of training material, the number of major concepts in any training session should be limited to seven, with each major concept being presented at least three times during the session. Trainees should be told at the outset of training what will be presented, why it is important, and how they should use it in their daily work activities. That information will convey the relevance of the training at the beginning of the session and prepare the learner.

The involvement of scientists in the development of content for health and safety training is particularly important. Often, employees who routinely care for nonhuman primates are not placed under the direct supervision of the scientist who is responsible for the research project, so there can be gaps in communication of information that is necessary to protect worker health and safety or information that could correct misperceptions about risks associated with the project. Such gaps could also place the research animals at unnecessary risk if exposed to a staff member who is an active carrier of an infectious agent. Safety-related information from the scientists needs to be conveyed both to the animal care staff and to their supervisors. Supervisors who understand the research objectives and the attendant hazards will create and maintain a safe work environment in which the animal care staff can be integral and knowledgeable participants in the research activity (NRC 1997).

The following are examples of the types of topics and content that may be presented in safety training:

- *Facility orientation.* Entrance requirements, facility policies, emergency procedures, and management overview.
- *Overview of the organization's safety program.* Integration of safety into SOP, the safety committee, safety elements of performance appraisals and position descriptions, and overview of the safety department and the organization's safety policies.
- *Hazards.* Hazards unique to nonhuman primates and the institution's specific activities.
- *Personal protective equipment.* Type required, function, protective capability, selection, individual fit, cleaning, storage, and replacement.
- *Personal protective clothing.* Function, protective capability, proper donning, wearing and removal, care, cleaning, and storage or disposal.
- *Eye protection.* Selection, individual variability, fit, task-specific selection, coverage, and care and storage.

- *First aid.* An overview of first aid to be administered in the event of an exposure.
- *Material safety data sheets.* What they are, how to use them, where they are stored, and limitations.
- *SOP.* What they are, how to use them, where they are stored, and limitations.
- *Species-specific behavior.* Threat behaviors, affiliative behaviors, individual differences, and descriptions of how the species responds to human behavior.
- *Sharps safety.* Proper selection, use, and disposal; proper use of safe sharp devices; and when to use these devices.
- *Protection of the back.* Proper lifting techniques and alternatives to lifting.
- *Safe movement of equipment.* Overview of equipment-related accidents and injuries, safe hand placement, hand protection, and overview of instances when more than one person may be needed to move a piece of equipment and why.

Selecting Methods and Techniques

Once the needs assessment has been done, the learning objectives developed, and the training content determined, training methods and techniques to accomplish the stated learning objectives can be selected. Types of training methods and techniques include traditional classroom style, on-the-job-training, practical hands-on training for groups and individuals, and computer-based training. For substantial learning to occur, learners must be actively involved in the learning process (Bird and Germain 1996). If the training pertains to humane restraint of an alert adult macaque, it may entail multiple methods, such as classroom instruction, practical exercise of restraint technique with a macaque model, and individual instruction with an animal.

Timing and Duration of Training

The time needed for instruction is based largely on the training objectives established and the content of the training. The more complex the issue, the more time may be required for training and learning. Generally speaking, more time should be allowed when conveying new information during training than for a refresher session or for teaching a variation of an existing operating standard.

Administration of Training

Trainees want to know why the information conveyed during training is important to them and their organization. They want to know how management expects them to use the new information. They want the training session to be organized and concise. And, they want to be treated with respect.

Five basic principles of adult learning have been identified (Bird and Germain 1996):

- *Principle of Readiness.* We learn best when we are ready to learn. You cannot teach someone something for which he or she does not have the necessary background of knowledge, maturity, or experience. Readiness also means that the learner is emotionally ready and motivated to learn; this readiness is created by communicating to the trainees why the training is important and how it will be of benefit to them (for example, in career progression, improved performance evaluations, and safer work).
- *Principle of Association.* It is easier to learn something new if it is built on something we already know. Start with the known practices and proceed to the new. Build up to the new information or more difficult tasks from the base of the known and familiar ones.
- *Principle of Involvement.* Substantial learning will occur when learners are actively involved in the learning process. The more senses involved (hearing, seeing, tasting, smelling, and feeling), the more effective the learning. Learner-involvement tools include practical or hands-on training, questioning of the group, open discussion, case studies and problem-solving, simulations, and demonstrations.
- *Principle of Repetition.* Repetition aids learning, retention, and recall. When there is a long period of disuse of information after it is learned, learned responses can weaken.
- *Principle of Reinforcement.* The more likely a response is to lead to satisfaction, the more likely it is to be learned and repeated. Instructors should use praise, reward, and recognition to reinforce learning.

Evaluating the Training

Learning is measured by a change in behavior. For example, are the trainees from the back-safety class demonstrating proper lifting techniques as demonstrated in class? Supervisors and managers must be knowledgeable about the content of training programs so that they can effectively monitor workplace practices and behavior to ensure adherence to prescribed methods or investigate and mitigate the reasons for discrepancies.

Several approaches can be used for evaluating the success of the training program. Periodic evaluation of the presentation, instructor, accommodations, and resources can assist in training program revision and improvement. That can be accomplished through student evaluation forms and monitoring of the training process by management and supervisors. Pre- and post-training testing can provide an indication of the information learned, how it is used, and changes in risk perception and attitudes. Other training-evaluation mechanisms include site inspection, personnel review, review of injury and illness records, review of regulatory-compliance citations, and periodic use of questionnaires. The approach should be carefully designed and applied to provide information useful for both institution officials and employees (NRC 1997).

One measure of training effectiveness is organizational impact. Has the organization experienced reduced costs associated with injuries since training inception? Perhaps there has been observable improvement in work quality or quantity. Improvements in worker-management relations can be effected through safety training. Safety should be integrated into the organization both horizontally and vertically, and safety-training efforts should be followed by careful review of activities and findings to develop recommendations for overall program improvement.

CONTINUING EDUCATION

Continuing education provides a broad background in which much of the information presented might not be directly related to the trainees' experience and needs. Training is job- and task-specific; training is directly related to performance and is intended to improve performance (Broadwell 1986). A qualifications assessment and a periodic individual needs assessment will assist those responsible for training to determine when continuing education, as opposed to training, is warranted. Does the person have an educational base that, when supplemented with the training that the institution is prepared to offer, will make it likely that the person will be qualified for a particular job? If the answer is no, continuing education is probably warranted. The resources required for offering continuing education are often different from those required for training, because of the broad background of the information to be presented. Institutions that use nonhuman primates for research, education, and testing often sponsor continuing education for assistant-technician, technician, and technologist certification examinations, and support veterinarians' preparation for board certification.

RECORDKEEPING

Recordkeeping is an essential aspect of a training program. Training records should reflect the date of the training, the instructor's name, names of attendees (documented with their initials), the location of the training, duration of training, broad topics, subtopics, learning objectives, course content, and a list of handout materials. Organizations such as AAALAC International and IACUCs may request review of training records during site visits and inspections. Some training records are required to be retained for specified durations to satisfy federal and state environmental health and safety regulations. Animals experimentally infected with HIV or HBV are included under the OSHA Bloodborne Pathogen Standard (29 CFR 1910.1030) in the category of "other potentially infectious materials." For training that is conducted to satisfy compliance with this standard, OSHA requires training records to be retained for 3 years. State OHS offices and regional OSHA representatives have information on specific requirements that may affect a facility. The institutional official is responsible for ensuring the maintenance of training records.

The head of the environmental health and safety office will usually strive to establish a simple system that presents the smallest administrative burden. A computer-based system should facilitate such an approach (NRC 1997).

9

Postexposure Medical Treatment in Nonhuman-Primate Facilities

Nonhuman primates are often ideal animal models for the study of human disease processes, but their phylogenetic similarity with humans places employees at risk for transmission of a broad variety of zoonoses as described in Chapter 3. Specific occupational health services recommendations for nonhuman-primate handlers have been presented in Chapter 7. This chapter addresses the role of an institution's occupational health care staff in prevention and treatment of staff exposure to nonhuman-primate related hazards.

Specific recommendations for occupational health services for nonhuman-primate handlers are outlined in *Biosafety in Microbiological and Biomedical Laboratories* (CDC-NIH 1999).

The occupational health services staff should work in concert with their facility's occupational health and safety program to ensure that workplace health risks are minimized and that appropriate medical services for evaluation and treatment are available to workers exposed to nonhuman primates or their tissues. These important occupational health services may be provided through a variety of arrangements depending on the individual facility risk-assessment process, as described in Chapter 5. It is critical that the designated medical providers be informed and appropriately updated regarding the nature of the health risks. That might require joint participation of research personnel, animal care staff, environmental health and safety staff, and occupational health services staff in continuing education regarding use of specific agents and safety training activities, as described in Chapters 7 and 8. Persons working with nonhuman primates or their tissues have the potential for exposure to infectious agents. Medical providers' participation in infection control and biosafety review may be appropriate and help to ensure that occupational health providers are updated on these important risks of infectious disease.

Many facilities are using emergency departments in local hospitals and clinics for obtaining emergency medical care. The individual responsible for the occupational health program should develop a close working relationship with the local emergency department to ensure that (1) emergency medical staff are familiar with the specific hazards posed by nonhuman primates and (2) they can provide a timely evaluation and follow up when an employee arrives at the emergency department. This may involve periodic meetings of the occupational health program and emergency department staff, training of the emergency department staff, and developing a packet of information for the emergency-room physician. This packet should outline concerns and offer information for telephone consultation with a knowledgeable physician.

DEFINING EXPOSURE RISK

The risk of developing infection from an exposure to nonhuman primates depends on many factors, including the nature of the infectious agent, the mechanism and route of exposure, the physiologic status of the source animal and the employee, whether personal protective equipment was used, and whether appropriate postexposure first-aid procedures were followed. Data that could be used to define risk for most exposure scenarios are sparse. Therefore, assessing the risks posed by some agents is problematic, and best estimates must be based on knowledge of similar agents or exposure situations. Health-care providers will have to approach employees' exposure concerns with a willingness to acknowledge unknowns while attempting to estimate risks and involve the employees in the clinical risk-assessment and decision tree process.

SCOPE OF POTENTIAL INFECTIOUS AGENTS IN NONHUMAN PRIMATES

The variety of potential infectious agents found in nonhuman primates is summarized in Chapter 3. Agents likely to be of concern in most facilities are reviewed here, but occupational health-care providers should become familiar with all exposures likely to be encountered in their facilities. For example, exposures to primates bred in captivity might pose different risks from exposures to wild-captured primates. Wild-captured primates can harbor naturally occurring zoonotic agents—including helminths, bacteria, and viruses—not typically found in captive-raised populations. Furthermore, experimental treatment of nonhuman primates with agents such as immunosuppressants can increase their risk for these infections and hence transmission to handlers.

In the handling of nonhuman primates, it is reasonable to adopt universal precautions and thereby minimize exposure risk regardless of the presence or level of infection (CDC-NIH 1999). When employee injuries occur, information on the viral status of the animal and any infectious experimental agents should be communicated to the patient's health professional. The patient should be managed on the basis of the assumption that the nonhuman primate involved was infected with the transmissible agent of concern. Defining the transmissible agent of concern is dependent on several factors: the species involved, the experimental status of the animal, the viral and microbiological status of the animal and/or colony.

Precautions must be taken in interpreting the risk posed by nonhuman primates classified as "specific pathogen free" animals because of the risk of latency of some infections and the fact that all diagnostic testing for the presence of infectious agents has technical limitations that lead to imperfect accuracy. B virus is well-known to become latent in macaques, and group-housed macaques can seroconvert and shed B virus without exhibiting clinical signs (Weigler and others 1993; Weir and others 1993). Additionally, due to the inaccuracy of some B virus tests, it should be assumed that all macaques, including those from SPF colonies, are infected (Ward and Hilliard 2002). Relying on past viral testing or serologic results does not provide adequate predictive information regarding viral shedding at the time of a human exposure to bites, scratches, or splashes from the animal. Although it should be assumed that all macaques are infected with B virus (even in SPF colonies) when initiating first aid and seeking medical care, decisions on antiviral prophylaxis treatment will include considerations of the viral status of the animal/colony and are discussed under "B Virus Exposure" later in this chapter.

DEFINING ROUTES OF EXPOSURE

The most common routes of exposure among nonhuman-primate handlers are scratches, needle sticks, cuts, bites, and mucous membrane exposures (bin Zakaria and others 1996), though most published reports of injury and infection among nonhuman-primate handlers are related to bites. Animal-bite wounds are more likely to result in bacterial infection than are simple abrasions or scratches because of the inherent limitations in cleansing of puncture wounds (Trott 1997). A review of bacterial infections after primate bites found that polymicrobial infections have often resulted in severe complications, including osteomyelitis caused by such bacterial flora as *Eikenella corrodens, Bacteroides* spp., and *Fusobacterium* spp. (Goldstein and others 1995).

Mucocutaneous exposures can also result in transmission of gastrointestinal pathogens, such as *Shigella* spp. (Kennedy and others 1993), *Campylobacter* spp. (Gibson 1998; Hubbard and others 1991; Johnson and others 2001; Renquist 1987; Taylor and others 1989; Tribe and Frank 1980), *Giardia intestinalis* (Armstrong and others 1979), *Crypotosporidium parvum* (Miller and others 1990; Muriuki and others 1998; Toft and Eberhard 1998), and *Entamoeba histolytica* (Remfry 1978).

Research involving retroviruses has demonstrated that SIV can infect humans. Occupational exposure to SIV has occurred through mucosal splashes, contamination of cuts or skin abrasions, and needle stick injuries (Essex 1994; Khabbaz and others 1994; Sotir and others 1997). The assumption that SIV infection can be transmitted by routes similar to those in occupational exposure to HIV has led the CDC to recommend that occupational exposures to SIV and hybrid strains of HIV-SIV be managed according to PHS postexposure prophylaxis guidelines for HIV (CDC 2001c).

Among the well-documented cases of human B virus infection, almost half involved exposure by monkey bite (Breen and others 1958; Bryan and others 1975; Cohen and others 2002; Davenport and others 1994; Davidson and Hummeler 1960; Holmes and others 1990; Hummeler and others 1959; Palmer 1987; Sabin and Wright 1934). Exposures also occurred by scratches (monkey or cage), mucosal exposures, exposures to blood or tissue (Cohen and others 2002) and by needle stick injury (Artenstein and others 1991). B virus can also be transmitted human-to-human by skin-to-skin contact (Holmes and others 1990).

While bites and scratches are of known importance in the transmission of B virus, systemic infection has also been reported in people not known to have been recently injured or exposed (Bryan and others 1975; Fierer and others 1973), suggesting that B virus latency in humans is a possibility. Because human herpesvirus latency in humans is known to exist, as is B virus latency in nonhuman primates, it might be presumed that B virus infection can become latent in humans (Fierer and others 1973). However, in a controlled seroprevalence study performed in 1995, there was no evidence of latent B virus infection in 321 handlers, using three high B virus antibody-titered humans who were available as references (Freifeld and others 1995). Additionally, other investigations have failed to document seroconversion among contacts of B virus-infected patients (Davenport and others 1994; Holmes and others 1990). However, several isolated cases in humans are suggestive that B virus latency in humans may occur (personal communication; Drs. Hilliard and Davenport). Current data would suggest that if B virus latency occurs at all in humans, it is very rare. The clinical significance of B virus latency in humans is even less well defined. It is unknown whether recurrent clinical infections or a chronic neurologic illness due to B virus occurs in humans. Therefore, clinicians responsible for evaluating injured or ill nonhuman-primate handlers should be aware that absence of recent known exposure to animals might not preclude B virus infection.

DETERMINING APPROPRIATE POSTEXPOSURE MEDICAL MANAGEMENT

As outlined above, human contact with nonhuman primates—including bites, scratches, and splashes—can lead to transmission of natural zoonoses or experimentally acquired infections, although the probability of infection is low under most circumstances. Although B virus infection occurs uncommonly after nonhuman-primate exposures, the illness is often fatal when left untreated, and so substantial care must be taken in assessing and following up on such contact (Holmes and others 1995). There should be follow-up monitoring of potential B virus exposure including tracking of any unexplained influenza-like illness lasting more than 48 hours in personnel working with nonhuman primates or their tissues, to allow for appropriate treatment advice by an occupational health physician.

Determination of appropriate postexposure medical management is not well defined, because the attendant risks are difficult to measure for each pathogen. Analogous occupational-exposure data do exist for some occupationally acquired infections, including HIV infection, in healthcare workers, but there is a paucity of data to support clinical decision-making in other settings. The medical care provider should rely on consensus documents and other guidelines, develop a network of infectious-disease specialists to assist in making treatment decisions, and be aware of the potential risks of iatrogenic injury or complications when making recommendations for postexposure management. For example, aggressive medical treatment might contribute to injury and illness through medication side effects if unnecessary prophylaxis is prescribed when exposure risks are uncertain or if patients are not monitored appropriately (CDC 2001a).

When assessing risk of infection after exposure to a nonhuman primate, the medical care provider should consider the risk factors associated with the nonhuman primate and the employee. This will necessitate close communication with the facility animal-care and infection-control staff. The initial communication should be used to gather information that will be important for decisions regarding prophylaxis and treatment and should include a discussion of the experimental, viral, and microbiological status of the animal and colony. For example, a nonhuman primate that is immunocompromised would potentially be a more likely source of B virus than otherwise (Chellman and others 1992; Holmes and others 1995). Similarly, circumstances regarding the kind of injury and the application of appropriate first aid should be taken into consideration in determining the risk of infection. In summary, defining the risk of infection after an exposure requires careful review of both the nonhuman primate and the employee in question.

FIRST AID AFTER EXPOSURES TO NONHUMAN PRIMATES

As with any human or animal exposure, postexposure first aid, including immediate cleansing of the wound or contaminated mucous membrane, is of the utmost importance in preventing infection. The importance of adequate wound cleaning must be stressed during employee training and should include a demonstration of methods and location in the workplace of cleansing materials. Nonhuman-primate facility directors and occupational health care providers should consider the use of a wallet card or other simple guide to remind employees about first aid and provide emergency contact information.

Good general guidelines for medical management of animal-bite wounds management are available (Smith and others 2000). In general, emergency first aid should be initiated before the patient travels to a health-care provider, because prompt cleaning may reduce the risk of infection with some pathogens. This is particularly important for injuries caused by macaques, because B virus becomes incorporated into host cells quickly (Ludwig and others 1983).

First aid for skin exposures should include flushing and scrubbing of the skin with a detergent and water for 15 minutes. Mechanical scrubbing of the wound margins is recommended in addition to the use of copious flowing water and detergent, povidone-iodine, or chlorhexidine (Cohen and others 2002) to further cleanse the wound of contamination. Exposed eyes or mucous membranes should be flushed for 15 minutes preferably at an eyewash station because it is difficult for employees to wash their own eyes. Detergents and antiseptics should *never* be used in the eyes. If an eye station is not in close proximity, workers should have ready access to a portable liter bag of sterile saline. Nonhuman-primate handling facilities must have appropriate eye-washing stations (in addition to standard sinks with running water) or appropriate eye-irrigation materials immediately available for first aid (ANSI standard Z358.1-1998). All nonhuman-primate facility employees must be taught about the location of sink and eye-washing facilities and the importance of early initiation of standard wound cleaning and mucous membrane irrigation.

MEDICAL EVALUATION AND FOLLOWUP

It is imperative that handlers of nonhuman primates have priority access to physicians or other appropriately trained medical staff who have specific knowledge and experience regarding injuries caused by nonhuman primates. It is the employee's responsibility to report an exposure or occurrence of symptoms. The employee's occupational health care provider is then responsible for assessment of the exposure or symptoms. It is imperative that the chain of command for report and assessment of an exposure or symptoms be clearly defined. This is particularly important for SIV and B virus exposures, for which prophylaxis, when advised, should be initiated as soon as possible to provide the greatest benefit. In addition to developing a local network, the medical care providers should have reference material available about unique aspects of exposure to nonhuman primates. Table 9-1, for example, provides guidelines for the initial management of exposures involving macaques.

A mechanism must be in place for reaching a knowledgeable healthcare provider outside ordinary working hours. For larger facilities, an on-call arrangement may be developed. For smaller facilities, the local emergency department may be utilized. Regardless of the size of the facility, the postexposure management plan must include training of employees and medical care providers. Regular follow-up must be provided after an injury; if an injured employee does not keep a recommended follow-up appointment, an effort must be made to inform the employee of the need for it.

Wound Assessment

Important factors to consider when assessing a wound caused by a nonhuman primate are outlined in Table 9-2 and should be documented in the medical record for later reference. The physical examination should include collection of vital signs and inspection of the wound or contaminated mucous membrane. This examination should be documented and should include information regarding the location, length, estimated depth, and shape of the wound, potential for nerve or tendon damage, vascular integrity, and presence of foreign body or contamination. First aid provided at the worksite should be documented. Animal bites and scratches pose a high risk of tetanus. Therefore, in addition to thorough cleaning, tetanus prophylaxis should be offered on the basis of CDC guidelines (CDC 1985).

TABLE 9-1 Initial Management of Exposures Involving Macaques

1. First Aid

   a. Mucous membrane exposures (eye, nose, or mouth):
      - Irrigation or flushing for 15 minutes with sterile saline solution or water. NEVER use detergent or disinfectants in the eye.

   b. Skin exposures:
      - Irrigation and scrubbing with soap or detergent solution and a high volume of water for 15 minutes. Use a detergent solution (such as chlorhexidine or povidone-iodine) if immediately available.

2. Initial Health Care Evaluation

   a. Patient:
      - Assess adequacy of first aid and consider additional wound cleaning.
      - Record mechanism of injury, likelihood of wound contamination, effectiveness of wound cleaning, tetanus status, and general evaluation of patient health (medication use, allergy to medications, and baseline medical condition).
      - Document examination of wound and overall physical examination sufficient to provide comparison if patient condition changes (dermatologic, ocular, respiratory, cardiovascular, lymph node, and neurologic examinations).
      - Consider collection of laboratory specimens, including wound cultures and blood sample.

   b. Primate:
      - Contact animal-care staff or infection-control staff to review health status of the colony, including consideration of infectious diseases (medication use, known experimental retrovirus infections, B virus serology status, and so on).
      - If an individual animal can be associated with an exposure, a clinical veterinarian experienced in nonhuman-primate medicine should evaluate the animal (physical examination; complete review of the medical, serologic, and research history; and blood draw and storage). In the case of potential B virus exposure, culture of the mucous membranes should be performed.

3. Risk Assessment

   a. On the basis of the above information, consider risk of infection or other illness.

   b. Counsel patient about the implications of the injury and the need for careful follow-up for wound evaluation and care. Provide antibiotic and/or antiviral prophylaxis if clinically indicated. In addition to scheduling a clinic appointment to re-evaluate the injury, give the patient emergency contact information.

TABLE 9-2 Wound Assessment

1. Circumstances of injury
   - Site of injury (for example, extremity, head, or neck)
   - Type of injury (bite, crush, scratch, or splash)
   - Species involved and infectious risk associated with it

2. Documentation of postexposure first aid

3. Past medical history
   - Tetanus immunization status
   - Medication allergies
   - Predisposing medical conditions (e.g.,immunocompromised, asplenic, or diabetic)

Risk of Bacterial Infection

Clinical data on which to base a decision to provide prophylactic antibiotic treatment for a nonhuman-primate bite are sparse. However, experience with bites of other animal species and humans suggests that the most important factors in preventing wound infection are related to initial wound cleaning. Wounds on the extremities, wounds that are macerated, or punctures are more likely to become infected than superficial, proximal wounds (Aghababian and Teuscher 1992; Goldstein 1992). Goldstein and others (1995) reviewed the subject of bacterial infection following nonhuman-primate bites and has advised that the diagnosis and treatment of simian bites should be similar to those of human bites. A rational approach to preventing bacterial infection would include empiric postbite antibiotic prophylaxis for all but the most superficial wounds. This should involve an antibiotic active against streptococci, staphylococci, enterococci, *Eikenella corrodens*, anaerobes, and *Enterobacteriaceae* spp. A good choice would be amoxicillin-clavulanic acid as a single agent or penicillin plus a penicillinase-resistant penicillin or cephalosporin. If the wound is not already infected, it is reasonable to initiate a 3- to 5-day course of treatment; if no signs of infection appear, the antibiotic treatment may be discontinued at the wound-check visit. Most physicians would not recommend primary closure of a human-bite wound on an extremity, but delayed closure could be considered during a follow-up visit if there is no sign of infection (Goldstein 2000).

B VIRUS EXPOSURE

Prophylaxis after B Virus Exposure

With respect to prophylaxis after exposure to B virus, a detailed description of postexposure management was issued in 1995 (Holmes and others). More recent guidelines suggest the following (Cohen and others 2002):

- Postexposure prophylaxis is *recommended* in the following circumstances: skin exposure with loss of skin integrity or mucosal exposure to a high-risk source (an ill macaque, an immunocompromised macaque, a macaque known to be shedding virus, or a macaque with lesions suggestive of B virus); an inadequately cleaned skin exposure with loss of skin integrity or mucosal exposure (with or without injury); head, neck, or torso laceration; deep puncture bite; needle stick associated with tissue or fluid from central nervous system; lesions suspicious for B virus on the animal's eyelids or mucosa; puncture or laceration after exposure to an object likely to be contaminated with fluid from oral or genital lesions, central nervous system tissues, or tissue known to contain virus; postcleaning culture positive for B virus; and immunocompromised status.
- Postexposure prophylaxis for B virus is *considered* in the following situations: mucosal or eye splash even if adequately cleaned; laceration with loss of skin integrity; needle stick with blood from ill or immunocompromised macaque; puncture or laceration after exposure to an object contaminated with body fluid (other than from a lesion); and a potentially infected cell culture.
- Postexposure prophylaxis for B virus is *not recommended* if the exposed skin remains intact or exposure was due to a nonhuman-primate species other than a macaque, unless such animal is highly likely to be ill from B virus due to rare cross-contamination occurrences from macaques (see Chapter 3).

Selecting Agent for Postexposure Prophylaxis

Acyclovir was suggested in the 1995 guidelines, to be given at 800 mg five times per day for 14 days (Holmes and other 1995). This regimen was based on animal studies demonstrating successful acyclovir treatment of experimental infection and case reports of successful acyclovir and gancyclovir treatment of B virus infection. Since those guidelines were published, valacyclovir has been released and has become the preferred antiviral agent because the active metabolite, acyclovir, is present at higher serum concentrations and probably leads to better compliance; it is given at 1 g three times per day for 2 weeks (Cohen and others 2002). Cohen and others (2002) also offer a detailed discussion of postexposure prophylaxis in pregnant women and those with renal insufficiency. Physicians should check for updates on the preferred treatment regimen following B virus exposure on the CDC and B virus laboratory web sites, since this report cannot stay up-to-date.

EXPOSURE TO SIMIAN IMMUNODEFICIENCY VIRUSES

SIVs were initially identified on the basis of their cross-reactivity with HIV in serum samples from rhesus macaques with AIDS-like illnesses (Kanki and others 1985). In 1994, Khabbaz and others published a case report of a laboratory worker who was accidentally infected with an SIV strain. This case led to a seroprevalence study of HIV2-SIV antibody among nonhuman-primate handlers and laboratory workers that demonstrated that 0.4% of handlers were seropositive (Sotir and others 1997). One of those seropositive workers demonstrated waning antibody titer over time; none developed any clinical manifestations of infection.

There are no postexposure-prophylaxis guidelines available for SIV, nor is there any known effective prophylactic treatment. The only postexposure-prophylaxis guidelines available for retrovirus exposures are for occupational exposures to HIV (CDC 2001c). The HIV guidelines advise the use of nucleoside reverse transcriptase inhibitors, sometimes in combination with a protease inhibitor. Treatment recommendations depend on the results of assessment of the circumstances of the exposure and on source and exposure factors. Factors to be considered include the type of exposure (percutaneous versus mucous membrane), type and amount of fluid (blood versus other fluids), infectious state of the source (stage of disease, viral load, and antiviral-treatment status), and susceptibility of the exposed person. However, postexposure prophylaxis for SIV exposures may be different because SIV may not have the same susceptibility to prophylaxis. The most recent guidance from the CDC HIV and Retrovirology Branch for SIV exposures is to consider two-drug treatment with nucleoside reverse transcriptase inhibitors after HIV exposure (CDC 2001c). Examples would include a 4-week treatment course with zidovudine and lamivudine or lamivudine and stavudine, depending on viral susceptibility. Initial medical assessment should include baseline serology, baseline complete blood count, chemistry panel, and pregnancy testing when appropriate. Follow-up medical assessment should include repeat clinical examination every 1-2 weeks during treatment. Serology should be assessed 6 weeks, 3 months, and 6 months after exposure (CDC-NIH 1999).

OTHER RETROVIRUSES

Simian foamy virus, simian T-lymphotropic virus, and simian D retrovirus infections are also studied in nonhuman primates. Handlers may be exposed to these retroviruses through contact with infected materials or injuries. Simian foamy virus infections have been documented among primate handlers (CDC 1997; Heneine and others 1998; Sandstrom and others 2000). Absence of documented secondary transmissions to close contacts or through the blood supply suggests that humans may be a dead-end host for this infection (Callahan and others 1999; Winkler and others 2000). Simian D retrovirus infection was also recently documented among nonhuman-primate handlers (Lerche and others 2001). In nonhuman primates, infection with these viruses has been associated with immunodeficiency syndromes and lymphoproliferative disorders.

RECOMBINANT-VACCINIA RESEARCH

Recent vaccine research has used live recombinant-vaccinia virus engineered to express antigens of herpesvirus, hepatitis B, rabies, influenza, and HIV among others (CDC 2001b). Recombinant-vaccinia research has led to great advances in the development of vaccines, including the development of an oral rabies vaccine for animals. However, a recent report of a human infection caused by a recombinant-vaccinia rabies virus (Rupprecht and others 2001) has raised concerns about the adequacy of health and safety recommendations for animal handlers who may be caring for nonhuman primates infected with vaccinia virus. The issue of vaccination of laboratory animal handlers has been re-examined, and revised guidelines are now available (CDC 2001b). The guidelines recommend that laboratory animal workers be vaccinated for vaccinia if they handle cultures or animals contaminated or infected with vaccinia virus, recombinant-vaccinia virus vectored vaccines, or other orthopoxviruses that infect humans (for example, monkeypox and cowpox). The special case of variola (smallpox) virus will not be described here.

References

Abbott, P.D. and K.S. Majeed. 1984. A survey of parasitic lesions in wild-caught, laboratory-maintained primates: (rhesus, cynomolgus, and baboon). Vet. Pathol. 21(2):198-207.

Aghababian, R.V. and J. Teuscher. 1992. Infectious diseases following major disasters. Ann. Emerg. Med. 21(4):362-367.

Aiello, S.E., ed. 1998. The Merck Veterinary Manual. 8th edition. Whitehouse Station, NJ: Merck & Co. Inc.

Alterman, L. 1995. Toxins and toothcombs: Potential allospecific chemical defenses in Nycticebus and Perodicticus. In: Creatures of the Dark: The Nocturnal Prosimians. L. Alterman, G.A. Doyle, and M.K. Izard, eds. New York: Plenum Press.

Aoyama, K., A. Ueda, F. Manda, T. Matsushira, T. Ueda, and C. Yamauchi. 1992. Allergy to laboratory animals: an epidemiological study. Br. J. Ind. Med. 49:41-47.

Armstrong, J., R.E. Hertzog, R.T. Hall, and G.L. Hoff. 1979. Giardiasis in apes and zoo attendants, Kansas City, Missouri. CDC Vet. Pub. Health Notes, January 1979.

Artenstein, A.W., C.B. Hicks, B.S. Goodwin, and J.K. Hilliard. 1991. Human infection with B virus following a needle stick injury. Rev. Infect Dis. 13(2):288-91.

ASME (American Society of Mechanical Engineers). 2001. ASME Boiler and Pressure Vessel Code. New York: ASME International.

Ator, N.A. 1991. Subjects and instrumentation. In: Techniques in the Behavioral and Neural Sciences: Experimental Analysis of Behavior, Part 1, I.H. Iversen, and K.A. Lattal, eds. Amsterdam: Elsevier Science Publishers.

Baker, H.J., L.G. Bradford, and L.F. Montes. 1971. Dermatophytosis due to *Microsporum canis* in a rhesus monkey. J. Am. Vet. Med. Assoc. 159:1607-1611

Barton, L.L. and M.B. Mets. 2001. Congential lymphocytic choriomeningitis virus infection: decade of rediscovery. Clin. Infect. Dis. 33:370-374

Battles, A.H., E.C. Greiner, and B.R. Collins. 1988. Efficacy of invermectin against natural infection of *Strongyloides* spp in squirrel monkeys (*Saimiri sciureus*). Lab. Anim. Sci. 38:474-476.

Bayne, K.A.L. and R. Davis. 1983. Susceptibility of the rhesus monkey to the Ponzo Illusion. Bull. Psychonomic Soc. 21(6):476-478.

Bayne, K.A.L., S.L. Dexter, J.K. Hurst, G.M. Strange, and E.E. Hill. 1993. Kong toys for laboratory primates: Are they really an enrichment or just fomites? Lab. Anim. Sci. 43(1):78-85.

Bennett, G.F. and W. McWilson. 1965. Transmission of a new strain of *Plasmodium cynomolgi* to man. J. Parasitol. 51:79-80.

Bernstein, I.S. 1991. Social housing of monkeys and apes: group formations. Lab. Anim. Sci. 41:329-333.

Bielli, M., S. Lauzi, A. Pratelli, M. Martini, P. Dall'Ara, and L. Bonizzi. 1999. Pseudotuberculosis in marmosets, tamarins, and Goeldi's monkeys (Callithrichidae/Callimiconidae) housed at a European zoo. J. Zoo Wildlife Med. 30(4):532-536.

bin Zakaria, M., N.W. Lerche, B.B. Chomel, and P.H. Kass. 1996. Accidental injuries associated with non-human primate exposure at two regional primate research centers (USA): 1988-1993. Lab. Anim. Sci. 146(3):298-304.

Bird, F.E. and G.L. Germain. 1996. Practical Loss Control Leadership. Loganville, GA: Net Norske Veritas, Inc.

Blosser, F. 1992. Primer on Occupational Safety and Health. Washington, DC: The Bureau of National Affairs.

Bower, W.A., O.V. Nainan, X. Han, and H.S. Margolis. 2000. Duration of viremia in hepatitis A virus infection. J. Infect. Dis. 182:12-17.

Brack, M. 1987. Agents transmissible from simians to man. Berlin: Springer-Verlag.

Brack, M. and F. Hosefelder. 1992. In vitro characteristics of *Yersinia pseudotuberculosis* of nonhuman primate origin. Zentralbl. Bakteriol. 277(3):280-287.

Breen, G.E., S.G. Lamb, A.T. Otaki. 1958. Monkey-bite encephalomyelitis: report of a case with recovery. Br. Med. J. 2:22-23.

Broadwell, M. M. 1986. The Supervisor and On-The Job Training. Reading, Mass: Addison-Wesley Publishing Company.

Brooks, J., E.W. Rud, R.G. Pilon, J.M. Smith, W.M. Switzer, and P.A. Sandstrom. 2002. Cross-species retroviral transmission from macaques to human beings. Lancet. 360:387-388.

Brown, D.W.G. 1997. Threat to humans from virus infections of non-human primates. Rev. Med. Virol. 7:239-246.

Bryan, B.L., C.D. Espana, R.W. Emmons, N. Vijayan, and P.D. Hoeprich. 1975. Recovery from encephalomyelitis caused by *Herpesvirus simiae*. Report of a case. Arch. Intern. Med. 135(6):868-870.

BLS (Bureau of Labor Statistics) US Department of Labor. 1986. Recordkeeping Guidelines for Occupational Injuries and Illness. Washington, DC: US Government Printing Office.

Callahan, M.E., W.M. Switzer, A.L. Matthews, B.D. Roberts, W. Heneine, T.M. Folks, and P.A. Sandstrom. 1999. Persistent zoonotic infection of a human with simian foamy virus in the absence of an intact orf-2 accessory gene. J. Virol. 73(1):9619-9624.

Carbone, M., P. Rizzo, and H. Pass. 2000. Simian virus 40: the link with human malignant mesothelioma is well established. Anticancer Res. 20(2A):875-877.

CDC (Centers for Disease Control and Prevention). 1985. Diphtheria, tetanus, and pertussis: guidelines for vaccine prophylaxis and other preventive measures. MMWR. 34(27):405-414, 419-426.

CDC (Centers for Disease Control and Prevention). 1987. Epidemiologic notes and reports of B virus infections in humans—Pensacola, Florida. MMWR. 36(19):289-290, 295-296.

CDC (Centers for Disease Control and Prevention). 1988. Perspectives in disease prevention and health promotion: guidelines to prevent simian immunodeficiency virus infection in laboratory workers and animal handlers. MMWR. 37(45):698-704.

CDC (Centers for Disease Control and Prevention). 1989. Ebola virus infection in imported primates—Virginia. MMWR. 38(48):831-832, 837-838.

CDC (Centers for Disease Control and Prevention). 1990a. Epidemiologic notes and reports update: Ebola-related filovirus infection in nonhuman primates and interim guidelines for handling nonhuman primates during transit and quarantine. MMWR. 39(2):22-24, 29-30.

CDC (Centers for Disease Control and Prevention). 1990b. Update: Filovirus infections among persons with occupational exposure to nonhuman primates. MMWR. 39(16):266, 273.

CDC (Centers for Disease Control and Prevention). 1991. Current trends update: nonhuman primate importation. MMWR. 40(40):684-685, 691.

CDC (Centers for Disease Control and Prevention). 1993. Tuberculosis in imported nonhuman primates—United States, June 1990—May 1993. MMWR. 42(29):572-576.

CDC (Centers for Disease Control and Prevention). 1996. Ebola-Reston virus infection among quarantined nonhuman primates—Texas, 1996. MMWR. 45(15):314-316.

CDC (Centers for Disease Control and Prevention). 1997. Nonhuman primate spumavirus infections among persons with occupational exposure - United States, 1996. MMWR. 46(96):129-131.

CDC (Centers for Disease Control and Prevention). 1998. Fatal Cercopithecine herpesvirus 1 (B virus) infection following mucocutaneous exposure and interim recommendations for worker protection. MMWR. 47(49):1073-1076, 1083.

CDC (Centers for Disease Control and Prevention). 2001a. Serious adverse events attributed to nevirapine regimens for postexposure prophylaxis after HIV exposures - worldwide. MMWR. 49(51):1153-1156.

CDC (Centers for Disease Control and Prevention). 2001b. Vaccinia (smallpox) vaccine recommendations of the Advisory Committee on Immunization Practices (ACIP). MMWR. 50(RR10):1-25.

CDC (Centers for Disease Control and Prevention). 2001c. Updated US Public Health Service guidelines for the management of occupational exposure to HBV, HCV and HIV and recommendations for post-exposure prophylaxis. MMWR. 50(RR11):1-42.

CDC-NIH (Centers for Disease Control and Prevention-National Institutes of Health). 1999. Biosafety in Microbiological and Biomedical Laboratories. HHS Publication No. (CDC) 93-8395, 4th ed. Washington, DC: U.S. Government Printing Office.

Chellman, G.J., V.S. Lukas, E.M. Eugui, K.P. Altera, S.J. Almquist, J.K. Hilliard. 1992. Activation of B virus (*Herpesvirus simiae*) in chronically immunosuppressed cynomologous monkeys. Lab. Animal Sci. 42(2):146-51.

Cohen, J.I., D.S. Davenport, J.A. Stewart, S. Deitchmann, J.K. Hilliard, L.E. Chapman and the B Virus Working Group. 2002. Recommendations for prevention of and therapy for exposure to B virus (Cercopithecine Herpesvirus 1). Clin. Infect. Dis.. 35:1191-1203.

Cohen, J.I., S. Feinstone, and R. H. Purcell. 1989. Hepatitis A virus infections in a chimpanzee: duration of viremia and detection of virus in saliva and throat swabs. J. Infect. Dis. 160:887-890.

Collins, W.E., L.H. Miller, R.H. Glew, P.G. Contacos, W.A. Howard, and J.D. Wyler. 1973. Transmission of three strains of *Plasmodium falciparum* from monkey to man. J. Parasitol. 59(5):855-858.

Dalgard, D.W., R.J. Hardy, S.L Pearson, G.J. Pucak, R.V. Quander, P.M. Zack, C.J. Peters, and P.B. Jahrling. 1992. Combined simian hemorrhagic fever and Ebola virus infection in cynomolgus monkeys. Lab. Anim. Sci. 42:152-157.

Davenport, D.S., D.R. Johnson, G.P. Homes, D.A. Jewett, S.C. Ross, and J.K. Hilliard. 1994. Diagnosis and management of human B virus (*Herpesvirus simiae*) infections in Michigan. Clin. Infect. Dis. 19(1):33-41.

Davidson, W.L. and K. Hummeler. 1960. B virus infection in man. Ann. NY Acad. Sci. 85:970-979.

Delong, D., L.W. Gerrity, and K. Bayne. 2001. Elements of an occupational health and safety program: deficiencies identified by AAALAC International. Lab. Anim. 30:(4)23-26.

Desem, D. and S.L. Jones. 1998. Development of a human gamma interferon enzyme immunoassay and comparison with tuberculin skin testing for detection of *Mycobacterium tuberculosis* infection. Clin. Diagn. Lab. Immunol. 5(4):531-536.

DeValois, R. L. and G.H. Jacobs. 1971. Vision. In: Behavior of Nonhuman Primates: Modern Research Trends. New York: Academic Press.

Dienstag, J.L., F.M. Davenport, R.W. McCollum, A.V. Hennessy, G. Klatskin, and R.H. Purcell. 1976. Nonhuman primate-associated viral hepatitis type A. J. Am. Med. Assoc. 236:462-464.

Douglas, J.D., K.N. Tanner, J.R. Prine, D.C. Van Ripper, and S.K. Derwelis. 1967. Molluscum contagiosum in chimpanzees. J. Am. Vet. Med. Assoc. 151:901-904.

Downie, A.W. 1974. Serological evidence of infection with Tana and Yaba pox viruses among several species of monkey. J. Hyg. Camb. 72:245-250.

Doyle, L., J. Markovits, and J. Roberts. 1988. Tularemia (*Francisella tularensis*) in a squirrel monkey (*Saimiri sciureus*). Lab. Anim. Sci. 38(4):491-492.

Dubois, A., D.E. Berg, N. Fiala, L.M. Herman-Ackah, G.I. Perez-Perez, and M.J. Blaser. 1998. Cure of *Helicobacter pylori* infection by omeprazole-clarithromycin-based therapy in non-human primates. J. Gastroenterol. 33:18-22.

Essex, M. 1994. Simian immunodeficiency virus in people. New Eng. J. Med. 330(3):209-210.

Favoretto, S.R., C.C. de Mattos, N.B. Morais, F.A.A. Araujo, and C.A. de Mattos. 2001. Rabies in marmosets (*Callithrix jacchus*), Ceara, Brazil. Emerg. Infect. Dis. 7(6):1062-1065.

Fear, F.A., M.E. Pinkerton, J.A. Cline, F. Kriewaldt, and S.S. Kalter. 1968. A leptospirosis outbreak in a baboon (*Papio* sp.) colony. Lab. Anim. Care. 18:22-28.

Ferber, D. 2002. Monkey virus link to cancer grows stronger. Science. 296:1012-1015.

Fierer, J., P. Bazely, and A.I. Braude. 1973. Herpes B virus encephalomyelitis presenting as ophthalmic zoster. A possible latent infection reactivated. Ann. Intern. Med. 79(2):225-8.

Formenty, P., C. Boesch, M. Wyers, C. Steiner, F. Donati, F. Dind, F. Walker, and B. Le Guenno. 1999. Ebola virus outbreak among wild chimpanzees living in a rain forest of Cote d'Ivoire. J. Infect. Dis. 179(Suppl 1):S120-126.

Fourie, P.B. and M.W. Odendaal. 1983. *Mycobacterium tuberculosis* in a closed colony of baboons (*Papio ursinus*). Lab. Anim. 17:125-128.

Fox, G.J., L. Handt, S. Xu, Z. Shen, E.F. Dewhurst, J.B. Paster, A.C. Dangler, K. Lodge, S. Motzel, and H. Klein. 2001. Novel *Helicobacter* species isolated from rhesus monkeys with chronic idiopathic colitis. J. Med. Microbiol. 50:421-429.

Freifeld, A.G., J. Hilliard, J. Southers, M. Muray, B. Savarese, J.M. Schmitt, and S.E. Straus. 1995. A controlled seroprevalence survey of primate handlers for evidence of asymptomatic herpes B virus infection. J. Infect. Dis. 171(4):1031-1034.

Fritz, J. 1994. Introducing unfamiliar chimpanzees to a group or partner. Lab. Primate Newsltr. 33:(1)5-7.

Fritz, P.E., J.G. Miller, M. Slayter, and T.J. Smith. 1986. Naturally occurring melioidosis in a colonized rhesus monkey (*Macaca mulatta*). Lab. Anim. 20:281-285.

Gallup, G.G. Jr. 1982. Self-awareness and the emergence of mind in primates. Am. J. Primatol. 2:237-248.

Gallup, G.G. Jr. 1977. Self-recognition in primates: a comparative approach to the bidirectional properties of consciousness. Am. Psychol. 32:329-338.

Gao, F., E. Bailes, D.L. Robertson, Y. Chen, C.M. Rodenburg, S.F. Michael, L.B. Cummins, L.O. Arthur, M. Peeters, G.M. Shaw, P.M. Sharp, and B.H. Hahn. 1999. Origin of HIV-1 in the chimpanzee *Pan troglodytes troglodytes*. Nature 397:436-441.

Georges-Courbot, M.C., A. Sanchez, C.Y. Lu, S. Baize, E. Leroy, J. Lansout-Soukate, C. Tevi-Benisan, A.J. Georges, S.G. Trappier, S.R. Zaki, R. Swanepoel, P.A. Leman, P.E. Rollin, C.J. Peters, S.T. Nichol, and T.G. Ksiazek. 1997. Isolation and phylogenetic characterization of Ebola viruses causing different outbreaks in Gabon. Emerg. Infect. Dis. 3(1):59-62.

Gershon, R.M. and B. G. Zirkin. 1995. Behavioral factors in safety training. In: Laboratory Safety: Principles and Practices. 2nd ed. D.O. Fleming, J. H. Richardson, and J. I. Tulis, eds. Washington, D.C.: American Society for Microbiology.

Ghandour, A.M., N.Z. Zahid, A.A. Banaja, K.B. Kamal, and A.I. Bouq. 1995. Zoonotic intestinal parasites of hamadryas baboons (*Papio hamadryas*) in the western and northern regions of Saudi Arabia. J. Trop. Med. Hyg. 98:431-439.

Gibson, S.V. 1998. Bacterial and mycotic diseases. In: Nonhuman Primates in Biomedical Research, Diseases. B.T. Bennett, C.R. Abee, and R. Henrickson, eds. San Diego, CA: Academic Press, Inc.

Goldman, L. and M. Feldman. 1949. Human infestation with scabies of monkeys. Arch. Dermatol. Syph. 59:175-178.

Goldstein, E.J. 1992. Bite wounds and infection. Clin. Infect. Dis. 14:633-638.

Goldstein, E.J. 2000. Bites. In Mandell, Douglas, and Bennett's Principles & Practice of Infectious Diseases. 5th ed. Philadelphia, PA: Churchill Livingstone.

Goldstein, E.J.C., E.P. Pryor, and D.M. Citron. 1995. Simian bites and bacterial infection. Clin. Infect. Dis. 20:1551-1552.

Gopal, R., A. Jeanes, M. Osman, C. Aylott, and J. Green. 2002. Marketing hand hygiene in hospitals—a case study. J. Hosp. Infect. 50(1):42-47.

Grandin, T. 1999. Safe Handling of Large Animals. In: Occupational Medicine: State of Art Reviews. R.L. Langley, ed. Philadelphia, PA: Hanley and Belfus, Inc.

Gugnani, H.C. 1971. *Trichophyton mentagrophytes* infection in monkeys and its transmission to man. Hind. Antibiot. Bull. 14:11-13.

Hallenbeck, W.H. 1993. Quantitative risk assessment for environmental and occupational health. 2nd ed. Boca Raton, Fl: Lewis Publishers.

Hamlen, H.J. and J.M. Lawrence. 1994. Giardiasis in laboratory-housed squirrel monkeys: a retrospective study. Lab. Anim. Sci. 44:235-239.

Heberling, R.L. and S.S. Kalter. 1971. Induction, course, and transmissibility of monkeypox in the baboon (*Papio cynocephalus*). J. Infect. Dis. 124:33-38.

Heckel, J.O., W. Rietschel, and F.T. Hufert. 2001. Prevalence of hepatitis B virus infections in nonhuman primates. J. Med. Primatol. 30(1):14-19.

Heneine, W., W.M. Switzer, P. Sandstrom, J. Brown, S. Vedapuri, C.A. Schable, A.S. Khan, N.W. Lerche, M. Schweizer, D. Neumann-Haefelin, L.E. Chapman, and T.M. Folks. 1998. Identification of a human population infected with simian foamy viruses. Nat. Med. 4:403-407.

Hershkovitz, P. 1975. Comments on the taxonomy of Brazilian marmosets (Callithris, Callitrichidae). Folia Primatol. (Basel) 24:137-172.

Hilliard, J.K. and R.D. Henkel. 1998. Unpublished data. Georgia State University: NIH B Virus Resource Laboratory, Viral Immunology Center.

Hirsch, V.M., R.A. Olmsted, M. Murphey-Corb, R.H. Purcell, and P.R. Johnson. 1989. An African primate lentivirus (SIVsm) closely related to HIV-2. Nature 339:389-392.

Ho, S.A., J.A. Hoyle, F.A. Lewis, A.D. Secker, D. Cross, N.P. Mapstone, M.F. Dixon, J.I. Wyatt, D.S. Tompkins, G.R. Taylor, and et al. 1991. Direct polymerase chain reaction test for detection of *Helicobacter pylori* in humans and animals. J. Clin. Microbiol. 29:2543-2549.

Holmes, G.P., J.K. Hilliard, K.C. Klontz, A.H. Rupert, C.M. Schindler, E. Parrish, D.G. Griffin, G.S. Ward, N.D. Bernstein, T.W. Bean, M.R. Ball, J.A. Brady, M.H. Wilder, and J.E. Kaplan. 1990. B virus (*Herpesvirus simiae*) infection in humans: epidemiologic investigation of a cluster. Ann. Intern. Med. 112(11):33-839.

Holmes, G.P., L.E. Chapman, J.A. Stewart, S.E. Straus, J.K. Hilliard, and D.S. Davenport. 1995. Guidelines for the prevention and treatment of B-virus infections in exposed persons. The B virus Working Group. Clin. Infect. Dis. 20:421-439.

Hubbard, G.B., D.R. Lee, and J.W. Eichberg. 1991. Diseases and pathology of chimpanzees at the Southwest Foundation for Biomedical Research. Am. J. Primatol. 24:273-282.

Hughes, R.A. and A.C. Keat. 1994. Reiter's syndrome and reactive arthritis: a current view. Semin. Arthritis Rheum. 24:190-210.

Hummeler, K, W.L. Davidson, W. Henle, A.C. LaBoccetta, and H.G. Ruch. 1959. Encephalomyelitis due to infection with *Herpesvirus simiae* (herpes B virus): a report of two fatal, laboratory-acquired cases. New Eng. J. Med. 261:64-68.

Hutin, Y.J., R.J. Williams, P. Malfait, R. Pebody, V.N. Loparev, S.L. Ropp, M. Rodriguez, J.C. Knight, F.K. Tshioko, A.S. Khan, M.V. Szczeniowski, and J.J. Esposito. 2001. Outbreak of human monkeypox, Democratic Republic of Congo, 1996-1997. Emerg. Infect. Dis. 7(3):434-438.

International Labour Office. 1998. J.M. Stellman. ed. Encyclopedia of Occupational Health and Safety. 4th ed. Geneva: International Labour Office.

Janda, H.D., H.D. Ringler, J.K. Hilliard, C.R. Hankin, and M.F. Hankin. 1990. Nonhuman primate bites. J. Orthopaed. Res. 8:146-150.

Jezek, Z., I. Arita, M. Szczeniowski, K.M. Pauluku, K. Ruti, and J.H. Nakano. 1985. Human tanapox in Zaire: clinical and epidemiological observations on cases confirmed by laboratory studies. Bull. World Health Organ. 63(6):1027-1035.

Johnson, D.L., M.L. Ausman, M.R. Rolland, V.L. Chalifoux, and G.R. Russell. 2001. Campylobacter-induced enteritis and diarrhea in captive cotton-top tamarins (*Saguinus oedipus*) during the first year of life. Comp. Med. 51(3):257-261.

Juan-Salles, C., J. Verges, and X. Valls. 1999. Shigellosis in a squirrel monkey: a clinical history. Vet. Rec. 145(528-529).

Kalter, S.S., C.H. Millstein, L.H. Boncyk, and L.B. Cummins. 1978. Tuberculosis in nonhuman primates as a threat to humans. Dev. Biol. Stand. 41:85-91.

Kanki, P.J., M.F. McLane, N.W. King, N.L. Letvin, R.D. Hunt, P. Sehgal, M.D. Daniel, R.C. Desrosiers, and M. Essex. 1985. Serologic identification and characterization of a macaque T-lymphotropic retrovirus closely related to HTLV-III. Science 228(4704): 1199-1201.

Katz, D., J.K. Hilliard, R. Eberle, S.L. Lipper. 1986. ELISA for detection of group-common and virus-specific antibodies in human and simian sera induced by herpes simplex and related simian viruses. J Virol Methods 14(2):99-109.

Kaufmann, A.F., J.I. Moulthrop, and R.M. Moore. 1975. A perspective of simian tuberculosis in the United States—1972. J. Med. Primatol. 4(5):278-286.

Kennedy, F.M., J. Astbury, J.R. Needham, and T. Cheasty. 1993. Shigellosis due to occupational contact with non-human primates. Epidemiol. Infect. 110(2):247-254.

Khabbaz, R.F., W. Heneine, J.R. George, B. Parekh, T. Rowe, T. Woods, W.M. Switzer, H.M. McClure, M. Murphey-Corb, and T.M. Folks. 1994. Brief report: infection of a laboratory worker with simian immunodeficiency virus. N. Engl. J. Med. 330(3):172-177.

Khodakevich, L., Z. Jezek, and K. Kinzanzka. 1986. Isolation of monkeypox virus from wild squirrel infected in nature. Lancet. 11;1(8472):98-99.

King, C.S., J.R. Hessler, and J.R. Broderson. 1999. Management of a biocontainment research Facility. Lab. Anim. 28(7):26-33.

Knezevich, M. 1998. Geophagy as a therapeutic mediator of endoparasitism in a free-ranging group of rhesus macaques (*Macaca mulatta*). Am. J. Primatol. 44:71-82.

Lambeth, S.P. and M.A. Bloomsmith. 1992. Mirror as enrichment for captive chimpanzees (*Pan troglodytes*). Lab. Anim. Sci. 42:261-266.

Lankas, G.R. and R.D. Jensen. 1987. Evidence of hepatitis A infection in immature rhesus monkeys. Vet. Pathol. 24:340-344.

Leary, G., K.A.L. Bayne, and C. Bennett. 1985. The ocular components of senescence. In: Behavior and Pathology of Aging in Rhesus Monkeys. R.T. Davis and C.W. Leathers, eds. New York: Alan R. Liss.

Lehner, N.L., M.J. Huerkamp, and D.L Dillehay. 1994. Reference Serum Revisited. Contemp. Topics in Lab. Anim. Sci. 33 (6): 61-63

Lerche, N.W., W.M. Switzer, J.L. Yee, V. Shanmugam, A.N. Rosenthal, L.E. Chapman, T.M. Folks, and W. Heneine. 2001. Evidence of infection with simian type D retrovirus in persons occupationally exposed to nonhuman primates. J. Virol. 75(4):1783-1789.

Levine, D.N. 1970. Protozoan parasites of nonhuman primates as zoonotic agents. Lab. Anim. Care. 20(2):377-382.

Levy D.A. and F. Leynadier. 2001. Latex allergy: review of recent advances. Curr. Allergy Rep. 1(1):32-8.

Loomis, M.R., T. O'Neill, M. Bush, and R.J. Montali. 1981. Fatal herpesvirus in patas monkeys and a black and white colobus monkey. J. Am. Vet. Med. Assoc. 179:1236-1239.

Ludwig, H., G. Pauli, H. Gelderblom, G. Darai, H. G. Koch, R. M. Flugel, B. Norrild, and M. D. Daniel. 1983. B-virus (*Herpesvirus simiae*). In: The Herpesviruses Vol. 2, B. Roizman, ed. New York: Plenum Press.

Mager, R.F. 1997. Preparing Instructional Objectives. Atlanta, GA: The Center for Effective Performance.

Malkin, D. 2002. Commentary: Simian virus 40 and non-Hodgkin lymphoma. Lancet 359: 812-813.

Mansfield, K. and N. King. 1998. Viral Diseases. In: Nonhuman Primates in Biomedical Research, Diseases. B.T. Bennett, C.R. Abee, and R. Henrickson, eds. San Diego, CA: Academic Press, Inc.

Mayhall, C.G., V.A. Lamb, and P.H. Coleman. 1981. Infection in rhesus (*Macaca mulatta*) and squirrel (*Saimiri sciureus*) monkeys due to *Mycobacterium tuberculosis* phage type B. Outbreak in a primate colony. J. Med. Primatol. 10:302-311.

McClure, H.M., B. Olberding, and L.M. Strozier. 1973. Disseminated herpesvirus infection in a rhesus monkey. J. Med. Primatol. 2:190-194.

Meiering, C.D. and M.L. Linial. 2001. Historical perspective of foamy virus epidemiology and infection. Clin. Microbiol. Rev. 14(1):165-176.

Menzel, E.W. and J. Lawson. 1985. Chimpanzee (*Pan troglodytes*) spatial problem solving with the use of mirrors and televised equivalents of mirrors. J. Comp. Psychol. 99:211-217.

Meyers, W.M., B.J. Gormus, G.P. Walsh, G.B. Basking, and G.B. Hubbard. 1991. Naturally acquired and experimental leprosy in nonhuman primates. Am. J. Trop. Med. Hyg. 44(4 Pt 2):24-27.

Michaels, M. 1998. Xenozoonoses and the xenotransplant recipient. Ann. NY Acad. Sci. 862:100-104.

Miller, R.A., M.A. Bronsdon, L. Kuller, and W.R. Morton. 1990. Clinical and parasitologic aspects of Cryptosporidiosis in nonhuman primates. Lab. Anim. Sci. 40:42-46.

Miranda, M.E., T.G. Ksiazek , T.J. Retuya, A.S. Khan, A. Sanchez, C.F. Fulhorst, P.E. Rollin, A.B. Calaor, D.L. Manalo, M.C. Roces, M.M. Dayrit, and C.J. Peters. 1999. Epidemiology of Ebola (subtype Reston) virus in the Philippines, 1996. J. Infect. Dis. 179(Suppl 1):S115-119.

Montali, R.J., E.C. Ramsey, C.B. Stephensen, M. Worley, J.A. Davis, and K.V. Homes. 1989. A new transmissible viral hepatitis of marmosets and tamarins. J. Infect. Dis. 160:759-765.

Most, H. 1973. *Plasmodium cynomolgi* malaria: accidental human infection. Am. J. Trop. Med. Hyg. 22(2):157-158.

Muchmore, E. 1975. Health program for people in close contact with nonhuman primates. In: M.L. Simmons, ed. Proceedings of NCI Symposium on Biohazards and Zoonotic Problems of Procurement, Quarantine, and Research. DHEW Publication No. 76-890.

Muchmore, E. 1987. An overview of biohazards associated with nonhuman primates. J. Med. Primatol. 16:55-82.

Munene, E., M. Otsyula, D.A.N. Mbaabu, W.I. Mutahi, S.M.R. Muriuki, and G.M. Muchemi. 1998. Helminth and protozoan gastrointestinal tract parasites in captive and wild-trapped African non-human primates. Vet. Parasitol. 78:195-201.

Muriuki, S.M.K., R.K. Murugu, E. Munene, G.M. Karere, and D.C. Chai. 1998. Some gastrointestinal parasites of zoonotic (public health) importance commonly observed in old world nonhuman primates in Kenya. Acta Tropica. 71:73-82.

Nakauchi, K. 1999. The prevalence of *Balantidium coli* infection in fifty-six mammalian species. J. Vet. Med. Sci. 61:63-65.

Ndao, M., N. Kelly, D. Normandin, J.D. Maclean, A. Whiteman, E. Kokoskin, I. Arevalo, and B.J. Ward. 2000. *Trypanosoma cruzi* infection of squirrel monkeys: comparison of blood smear examination, commercial enzyme-linked immunosorbent assay, and polymerase chain reaction analysis as screening tests for evaluation of monkey-related injuries. Comp. Med. 50:658-665.

NIH (National Institutes of Health). 1998. Guidelines for Research Involving Recombinant DNA Molecules (NIH Guidelines). Rockville, MD: National Institutes of Health.

NIH (National Institutes of Health) and Occupational Medicine Service. 2001. Animal exposure surveillance program. Rockville, MD: National Institutes of Health.

NIOSH (National Institute of Occupational Safety and Health). 1977. Criteria for a recommended standard: occupational exposure to waste anesthetic gases and vapors. NIOSH Publication 77-140. Washington, DC: U.S. Government Printing Office.

NIOSH (National Institute of Occupational Safety and Health). 1999. Cercopithecine herpesvirus I (B virus) infection resulting from ocular exposure, NIOSH Publication 99-100. Washington, DC: U.S. Government Printing Office.

Nowak, R. M. 1999. Walker's Primates of the World. Baltimore, MD: The Johns Hopkins University Press.

NRC (National Research Council). 1983. Risk Assessment in the Federal Government: Managing the Process. Washington, DC: National Academy Press.

NRC (National Research Council). 1995. Prudent Practices in the Laboratory: Handling and Disposal of Chemicals. Washington, DC: National Academy Press.

NRC (National Research Council) and Institute for Laboratory Animal Resources Committee to Revise the Guide. 1996. Guide for the Care and Use of Laboratory Animals. 7th ed. Washington, DC: National Academy Press.

NRC (National Research Council). 1997. Occupational Health and Safety in the Care and Use of Research Animals. Washington, DC: National Academy Press.

NRC (National Research Council) and Committee on Well-Being of Nonhuman Primates. 1998. The Psychological Well-Being of Nonhuman Primates. Washington, DC: National Academy Press.

NRC (National Research Council) and Committee on Guidelines for the Use of Animals in Neuroscience and Behavioral Research. In press. Guidelines for the Care and Use of Animals in Neuroscience and Behavioral Research. Washington, DC: The National Academies Press.

NSC (National Safety Council). 1996. Fundamentals of Industrial Hygiene. 4th ed. Itasca, IL: National Safety Council.

NSC (National Safety Council). 1997. G.R. Krieger and J.F. Montgomery. eds. Accident Prevention Manual for Business and Industry: Administration and Programs. 11th ed. Itasca, Ill: National Safety Council.

Oberste, M.S., K. Maher, and M.A. Pallansch. 2002. Molecular phylogeny and proposed classification of the simian picornaviruses. J. Virol. 76:1244-1251.

OLAW (Office for Laboratory Animal Welfare), National Institutes of Health. 2000. Public Health Service Policy on Humane Care and Use of Laboratory Animals. Rockville, MD: National Institutes of Health.

Ollomo, B., S. Karch, P. Bureau, N. Elissa, A.J. Georges, and P. Millet. 1997. Lack of malaria parasite transmission between apes and humans in Gabon. Am. J. Trop. Med. Hyg. 56(4):440-445.

Osborne, C.G., M.D. McElvaine, A.S. Ahl, and J.W. Glosser. 1995. Risk analysis systems for veterinary biologics: a regulator's tool box. Rev. Sci. Tech. Off. Int. Epiz. 14(4):925-935.

OSHA (Occupational Safety and Health Administration). 1991. Preamble Section V. Quantitative risk assessment. Occupational Exposure to Bloodborne Pathogens. Fed. Reg. 56:64004-64182.

OSHA (Occupational Safety and Health Administration). 1999. Record summary of the request for information on occupational exposure to blood-borne pathogens due to percutaneous injury. www.osha-slc.gov/html/ndlreport052099.html.

Palmer, A.E. B virus, *Herpesvirus simiae*: historical perspective. J. Med. Primatol. 16: 99-130, 1987.

Perolat, P., J.P. Poingt, J.C. Vie, C. Jouaneau, G. Baranton, and J. Gysin. 1992. Occurrence of severe leptospirosis in a breeding colony of squirrel monkeys. Am. J. Trop. Med. Hyg. 46(5):538-545.

Penner, L. R. 1981. Concerning threadworm *(Strongyloides stercoralis)* in great apes - lowland gorillas and chimpanzees. J. Zoo Anim. Med. 12(4):128-131.

Petry, R.W., M.J. Voss, L.A. Kroutil, W. Crowley, R.K. Bush, and W.W. Busse. 1985. Monkey dander asthma. J. Allergy Clin. Immun. 75(2):268-271.

Phillips-Conroy, J.E., C.J. Jolly, B. Petros, J.S. Allan, and R.C. Desrosiers. 1994. Sexual transmission of SIVagm in wild grivet monkeys. J. Med. Primatol. 23:1-7.

Plesker, R. and M. Claros. 1992. A spontaneous *Yersinia pseudotuberculosis* infection in a monkey colony. Zentralbl. Veterinarmed. [B]. 30:201-208.

Poole, A.G., S.M. Shane, M.T. Kearney, and D.A. McConnell. 1999. Survey of occupational hazards in large animal practices. J. Am. Vet. Med. Assoc. 215(10):1433-5.

Poole, A.G., S.M. Shane, M.T. Kearney, and W. Rehn. 1998. Survey of occupational hazards in companion animal practices. J. Am. Vet. Med. Assoc. 212(9):1386-8.

Poyry, T., L. Kinnunen, T. Hovi, and T. Hyypia. 1999. Relationships between simian and human enteroviruses. J. Gen. Virol. 80:635-638.

Purcell, R.H. and S.U. Emerson. 2001. Animal models of hepatitis A and E. ILAR J. 42(2):161-177.

Rayan, G.M., D.J. Flournoy, and S.L. Cahill. 1987. Aerobic mouth flora of the rhesus monkey. J. Hand Surg - Am. 12A:299-301.

Rayburn, S.R. 1990. Special Medical Factors. In: The Foundation of Laboratory Safety: A Guide for the Biomedical Laboratory. New York, Berlin, Heidelberg: Springer-Verlag.

Reindel, J.F., A.L. Fitzgerald, M.A. Breider, A.W. Gough, C. Yan, J.V. Mysore, and A. Dubois. 1999. An epizootic of lymphoplasmacytic gastritis attributed to *Helicobacter pylori* infection in cynomolgus monkeys (*Macaca fascicularis*). Vet. Pathol. 36:1-13.

Reinhardt, V. 1991. Group formation of previously single-caged adult rhesus macaques for the purpose of environmental enrichment. J. Exp. Anim. Sci. 34:110-115.

Reinhardt. V. 1988. Preliminary comments on pairing unfamiliar adult male rhesus monkeys for the purpose of environmental enrichment. Lab. Primate Newsltr. 27(4):1-3.

Reinhardt, V. 1990. Social enrichment for laboratory primates: a critical review. Lab. Primate Newsltr. 29(3):7-11.

Remfry, J. 1978. The incidence, pathogenesis and treatment of helminth infections in rhesus monkeys (*Macaca mulatta*). Lab Anim. 12(4):213-8.

Renquist, D.M. 1987. Selected biohazards of naturally infected nonhuman primates. J. Med. Primatol. 16:91-97.

Richardson, J.H. 1987. Basic considerations in assessing and preventing occupational infections in personnel working with nonhuman primates. J. Med. Primatol. 16:83-89.

Richardson, J.H. and G.L. Humphrey. 1971. Rabies in imported nonhuman primates. Lab. Anim. Sci. 21:1082-1083.

Robertson, B.H. 2001. Viral hepatitis and primates: historical and molecular analysis of human and nonhuman primate hepatitis A, B, and the GB-related viruses. J. Viral Hepatitis 8:233-242.

Robertson, B.H. and H.S. Margolis. 2002. Primate hepatitis B viruses – genetic diversity, geography, and evolution. Rev. Med. Virol. 12:133-141.

Robertson, L. S. 1998. Injury Epidemiology Research and Control Strategies. 2nd ed. New York, New York: Oxford University Press, Inc.

Rollin, P.E., R.J. Williams, D.S. Bressler, S. Pearson, M. Cottingham, G. Pucak, A. Sanchez, S.G. Trappier, R.L. Peters, P.W. Greer, S. Zaki, T. Demarcus, K. Hendricks, M. Kelley, D. Simpson, T.W. Geisbert, P.B. Jahrling, C.J. Peters, and T.G. Ksiazek. 1999. Ebola (subtype Reston) virus among quarantined nonhuman primates recently imported from the Philippines to the United States. J. Infect. Dis. 179(Suppl 1):S108-114.

Ronald, N.C. and J.E. Wagner. 1973. Pediculosis of spider monkeys: a case report with zoonotic implications. Lab. Anim. Sci. 23(6):872-875.

Rupprecht, C.E., L. Blass, K. Smith, L.A. Orciari, M. Niezgoda, S.C. Whitfield, R.V. Gibbons, M. Guerra, and C.A. Hanlon. 2001. Human infection due to recombinant vaccinia-rabies glycoprotein virus. N. Engl. J. Med. 23 (Aug):345-348, 582-586.

Sabin, A.B. and Wright, A.M. 1934. Acute ascending myelitis following a monkey bite, with the isolation of a virus capable of reproducing the disease. J. Exp. Med. 59:115-136.

Samet, J. M. and T.A. Burke. 1998. Epidemiology and risk assessment. In: Applied Epidemiology. Theory to Practice. R.C. Brownson and D.B. Petitti, eds. New York: Oxford University Press.

Sandstrom, P.A., K.O. Phan, W.M. Switzer, T. Fredeking, L. Chapman, W. Heneine, and T.M. Folks. 2000. Simian foamy virus infection among zoo keepers. Lancet 355:551-552.

Santiago, M.L., C.M. Rodenburg, S. Kamenya, F. Bibollet-Ruche, F. Gao, E. Bailes, S. Meleth, S.-J. Soong, J.M. Kilby, Z. Moldoveanu, B. Fahey, M.N. Muller, A. Ayouba, E. Nerrienet, H.M. McClure, J.L. Heeney, A.E. Pusey, D.A. Collins, C. Boesch, R.W. Wrangham, J. Goodall, P.M. Sharp, G.M. Shaw, and B.H. Hahn. 2002. SIVcpz in wild chimpanzees. Science. 295: 465.

Sapolsky, R.M. and J.G. Else. 1987. Bovine tuberculosis in a wild baboon population: epidemiological aspects. J. Med. Primatol. 16:229-235.

Sargeaunt, P.G., J.E. Williams, and D.M. Jones. 1982. Electrophoretic isoenzyme patterns in *Entamoeba histolytica* and *Entamoeba chattoni* in a primate survey. J. Protozool. 29(1):136-139.

Schoeb, T.R., K. Dybvig, K.F. Keisling, M.K. Davidson, and J.K. Davis. 1997. Detection of urogenital mycoplasmal infections in primates by use of polymerase chain reaction. Lab. Anim. Sci. 47(5):468-471.

Schou, S. and A.K. Hansen. 2000. Marburg and Ebola virus infections in laboratory nonhuman primates: a literature review. Comp. Med. 50:108-123.

Scinicariello, F., R. Eberle, J.K. Hilliard. 1993. Rapid detection of B virus (*herpesvirus simiae*) DNA by polymerase chain reaction. J Infect Dis. 168(3):747-750.

Sherman, R.A. and K.J. Shimoda. 2001. Tuberculosis tracking: determining the frequency of the booster effect in patients and staff. Am. J. Infect.Control 29(1):7-12.

Sibal, L.R. and K.J. Samson. 2001. Nonhuman primates: A critical role in current disease research. ILAR J. 42:74-84.

Smith, J.M. and E. Meerovitch. 1985. Primates as a source of *Entamoeba histolytica*, their zymodeme status and zoonotic potential. J. Parasitol. 71:751-756.

Smith, P.F., A.M. Meadowcroft, and D.B. May. 2000. Treating mammalian bite wounds. J. Clin. Pharm. Therm. 25:85-99.

Solomon, T. and M. Mallewa. 2001. Dengue and other emerging flaviviruses. J. Infect. 42:104-115.

Sotir, M., W. Switzer, C. Schable, J. Schmitt, C. Vitek, and R.F. Khabbaz. 1997. Risk of occupational exposure to potentially infectious nonhuman primate materials and to simian immunodeficiency virus. J. Med. Primatol. 26:233-240.

Stephensen, C.B., J.R. Jacob, R.J. Montali, K.V. Holmes, E. Muchmore, R.W. Compans, E.D. Arms, M.J. Buchmeier, and R.E. Lanford. 1991. Isolation of an arenavirus from a marmoset with callitrichid hepatitis and its serological association with disease. J. Virol. 65:3995-4000.

Strickler, H.D. and J.J. Goedert. 1998. Exposure to SV40-contaminated poliovirus vaccine and the risk of cancer—a review of the epidemiological evidence. Dev. Biol. 94:235-244.

Szeinuk, J., W. S. Beckett, N. Clark, and W.L. Hailoo. 2000. Medical evaluation for respirator use. Am. J. Ind. Med. 32(1):142-157.

Takahashi, K., B. Brotman, S. Usuda, S. Mishiro, and A.M. Prince. 2000. Full-genome sequence analyses of hepatitis B virus (HBV) strains recovered from chimpanzees infected in the wild: implications for an origin of HBV. Virol. 267:58-64.

Takasaka, M., A. Kohno, I. Sakakibara, H. Narita, and S. Honjo. 1988. An outbreak of salmonellosis in newly imported cynomolgus monkeys. Jpn. J. Med. Sci. Biol. 41(1):1-13.

Taylor, N.S., M.A. Ellenberger, P.Y. Wu, and J.G. Fox. 1989. Diversity of serotypes of *Campylobacter jejuni* and *Campylobacter coli* isolated in laboratory animals. Lab. Anim. Sci. 39:219-222.

Thompson, S.A., J.K. Hilliard, D. Kittel, S. Lipper, W.E. Giddens, D.H. Black, and R. Eberle. 2000. Retrospective analysis of an outbreak of B virus infection in a colony of DeBrazza's monkeys (*Cercopithecus neglectus*). Comp. Med. 50:649-657.

Toft, J.D. and M.L. Eberhard. 1998. Parasitic diseases. In: Nonhuman Primates in Biomedical Research, Diseases. B.T. Bennett, C.R. Abee, and R. Henrickson, eds. San Diego, CA: Academic Press, Inc.

Trakulsomboon, S., T.L. Pitt, and D.A.B. Dance. 1994. Molecular typing of *Pseudomonas pseudomallei* from imported primates in Britain. Vet. Record. 135:65-66.

Tribe, G.W. and A. Frank. 1980. Campylobacter in monkeys. Vet. Record. 106:365-366.

Tribe, G.W. and E. Noren. 1983. Incidence of bites from cynomolgus monkeys in attending animal staff—1975-80. Lab. Anim. 17:110.

Trott, A.T. 1997. Wounds and Lacerations: Emergency Care and Closure. 2nd ed. St. Louis, MO: Mosby-Year Book, Inc.

Valverde, C.R., D. Canfield, R. Tarara, M.I. Esteves, and B.J. Gormus. 1998. Spontaneous leprosy in a wild-caught cynomolgus monkey. Int. J. Lepr. Other Mycobact. Dis. 66:140-148.

Vilchez, A.R., R.C. Madden, A.C. Kozinetz, J.S. Halvorson, S.Z. White, L.J. Jorgensen, J.C. Finch, and S.J. Butel. 2002. Association between simian virus 40 and non-Hodgkin lymphoma. Lancet 359:817-823.

Vore, S.J., P.D. Peele, P.A. Barrow, J.F. Bradfield, and W.H. Pryor. 2001. A prevalence survey for zoonotic enteric bacteria in a research monkey colony with specific emphasis on the occurrence of enteric Yersini. J. Med. Primatol. 30:20-25.

Ward, J.A. and J.K. Hilliard. 2002. Herpes B virus specific pathogen-free breeding colonies of macaques (*Macaca mulatta*): serological test results and the B virus status of the macaque. Contemp. Topics in Lab. Anim. Sci. 41(4):36-41.

Ward, J.A., J.K. Hilliard, and S. Pearson. 2000. Herpes B-virus specific-pathogen-free breeding colonies of macaques (*Macaca mulatta*): diagnostic testing before and after elimination of the infection. Comp. Med. 50:317-322.

Washburn, D.A. and D.M. Rumbaugh. 1992. Investigations of rhesus monkey video task performance: evidence for enrichment. Contemp. Topics in Lab. Anim. Sci. 31(5):6-10.

Weber, H., E. Berge, J. Finch, P. Heidt, F.-J. Kaup, G. Perretta, B. Verschuere, and S. Wolfensohn. 1999. Health monitoring of non-human primate colonies. Lab. Anim. 33 (Suppl.1):S1:3-S1:18.

Weigler, B.J. 1992. Biology of B virus in macaque and human hosts: a review. Clin. Infect. Dis. 14:555-567.

Weigler, B.J., D.W. Hird, J.K. Hilliard, N.W. Lerche, J.A. Roberts, and L.M. Scott. 1993. Epidemiology of cercopithecine herpesvirus 1 (B virus) infection and shedding in a large breeding cohort of rhesus macaques. J Infect Dis. 167(2):257-263.

Weigler, B.J., E. Scinicariello, and J.K. Hilliard. 1995. Risk of venereal B virus (cercopithecine herpesvirus 1) transmission in rhesus monkeys using molecular epidemiology. J. Infect. Dis. 171:1139-1143.

Weigler, B.J., J.A. Roberts, D.W. Hird, N.W. Lerche, and J.K. Hilliard. 1990. A cross-sectional survey for B virus antibody in a colony of group housed rhesus macaques. Lab. Anim. Sci. 40(3):257-261.

Weigler, B.J. and R.A. Ponce. 1999. A quantitative risk model for laboratory-acquired Herpes B virus infection at a regional primate research facility. Infect. Dis. Rev. 1(3):200.

Weir, E.C., P.N. Bhatt, R.O. Jacoby, J.K. Hilliard, and S. Morgenstern. 1993. Infrequent shedding and transmission of *herpesvirus simiae* from seropositive macaques. Lab. Anim. Sci. 43(6):541-544.

Whitley, R.J. and J.K. Hilliard. 2001. Cercopithecine herpesvirus 1 (B virus). In: Fields Virology. 4th ed. D.M. Knipe and P.M. Howley, eds. Philadelphia, PA: Lippincott Williams & Wilkins.

Whitney, R.A. Jr. 1976. Important primate diseases (biohazards and zoonoses). Cancer Res. Saf. Monogr. 2:23-52.

Wilson, R.B., M.A. Holscher, T. Chang, and J.R. Hodges. 1990. Fatal *herpesvirus simiae* (B virus) infection in a pata monkey (*Erythrocebus patas*). J. Vet. Diagn. Invest. 2:242-244.

Winkler, I.G., R.M. Flugel, K. Asikainen, and R.L.P. Flower. 2000. Antibody to human foamy virus not detected in individuals treated with blood products or in blood donors. Vox Sang. 79:118.

Wolfenshon, S. 1998. Shigella infection in macaque colonies: case report of an eradication and control program. Lab. Anim. Sci. 48:330-333.

Zumpe, D., M.S. Siberman, and R.P. Michael. 1980. Unusual outbreak of tuberculosis due to *Mycobacterium bovis* in a closed colony of rhesus monkeys (*Macaca mulatta*). Lab. Anim. Sci. 30(2 Pt. 1):237-240.

Appendix A

Workshop Speakers

Christian Abee, DVM
Department of Comparative Medicine
University of South Alabama

Thomas M. Butler, DVM
Department of Laboratory Animal Medicine
Southwest Foundation for Biomedical Research

Louisa Chapman, MD
National Center for Infectious Diseases
Centers for Disease Control and Prevention

Jon Crane
CUH2A, Inc.

David Davenport, MD
College of Human Medicine
Michigan State University

Thomas DeMarcus
National Center for Infectious Diseases
Centers for Disease Control and Prevention

Peter J. Gerone, Sc.D.
Tulane Regional Primate Research Center

Thomas Gordon
Yerkes Regional Primate Research Center
Emory University

James M. Schmitt, MD
Occupational Medical Service, DS
National Institutes of Health

Deborah Wilson, PhD
Occupational Safety and Health Branch
National Institutes of Health

Melinda Young, CBSP
Washington National Primate Research Center

Appendix B

Committee Member Biographies

Frederick A. Murphy, D.V.M., Ph.D. (IOM) is Professor of Virology and Dean Emeritus of the University of California, Davis, School of Veterinary Medicine. His expertise is in virology, veterinary medicine in general, and in the area of new and emerging zoonotic diseases in particular. As an IOM member, he has served on numerous IOM committees, including the Committee on Emerging Microbial Threats to Health in the 21st Century.

Jeffrey A. Roberts, D.V.M. is the Assistant Director of Primate Services at the California National Primate Research Center in Davis. He recently directed a related study for the national primate research centers. He has extensive experience as a primate veterinarian, and has been involved in several workshops and studies in the area of occupational health and safety in the care of nonhuman primates.

Kathryn A. L. Bayne, M.S., Ph.D., D.V.M. is Associate Director of the Association for Assessment and Accreditation of Laboratory Animal Care International (AAALAC). Her expertise is in nonhuman primate behavior and in the evaluation of laboratory animal care programs, including institutional occupational health and safety programs related to the use of animals. She has served on two NRC committees, the Committee to Revise the Guide for the Care and Use of Laboratory Animals and the Committee on The Psychological Well-Being of Nonhuman Primates, and is currently on the Committee on Guidelines for the Care and Use of Mammals in Neuroscience and Behavioral Research.

James L. Blanchard, D.V.M., Ph.D. is Associate Director for Veterinary Resources and the Head of the Department of Veterinary Medicine at the Tulane Regional Primate Research Center in Covington, LA. He is a primate veterinarian and has broad experience in parasitology and in occupational health and safety issues in nonhuman primate facilities.

Thomas J. Ferguson, M.D., Ph.D. is the Director of Student Health Services at the University of California, Davis. His expertise is in occupational medicine having been responsible for clinical occupational health services at UC Davis including the management of workplace exposures to nonhuman primates. He is board certified in internal medicine, occupational medicine, and medical toxicology and also holds a Ph.D. degree in public health with emphasis in environmental health sciences (UCLA).

LCDR Lisa J. Flynn, M.S. is an Environmental Health Specialist in the Office of the Commissioner, US Food and Drug Administration. Her expertise is in occupational health and safety issues. She spent 12 years at NIH first as Occupational Health and Safety Specialist and then as a Program Manager. During her tenure there, she participated in the development of the NIH Manual Issuance on the protection of personnel working with nonhuman primates.

Jack O. Geissert, M.P.H., C.I.H. is the Director of Environmental Health and Safety for the Massachusetts operations of Genetics Institute-Wyeth Pharmaceuticals. He is a certified industrial hygienist. He has served at numerous institutions in this capacity, including NIOSH, Colorado State University, and several corporations.

Julia K. Hilliard, Ph.D. is a Georgia Research Alliance Eminent Scholar and Professor and Director of the National B Virus Resource Center within the Viral Immunology Center in the Department of Biology at Georgia State University. She is one of the world's pre-eminent experts on the biology and transmission of B virus. She previously served on the NRC Committee on Occupational Safety and Health in Research Animal Facilities.

Michael P. Kiley, Ph.D. is Research Programs Safety Officer for the Agricultural Research Service, US Department of Agriculture. His expertise is in developing and overseeing multiple levels of biocontainment facilities.

Clarence J. Peters, M.D. is Professor in the Department of Microbiology and Immunology and Pathology at the University of Texas Medical Branch (UTMB) in Galveston. He is an expert in the area of public health and infectious disease and spent eight years as Chief of the Special Pathogens Branch, Division of Viral and Rickettsial Diseases at the National Center for Infectious Diseases, CDC, prior to moving to UTMB.

Benjamin J. Weigler, D.V.M., M.P.H., Ph.D. is the Director of Animal Health Resources and Attending Veterinarian at the Fred Hutchinson Cancer Research Center, Seattle, WA. His specialties are laboratory animal medicine, epidemiology of B virus transmission and elimination from colonies of nonhuman primates, biostatistics, herpesviruses, and zoonotic diseases. His expertise is in epidemiology and risk assessment and in addressing occupational health concerns in the research animal workplace.

Consultant

David S. Davenport, M.D., F.A.C.P. is an Associate Professor of Medicine at Michigan State University College of Human Medicine. He is board certified in internal medicine and infectious diseases. His expertise is in the diagnosis and management of B virus infections in humans. He has been involved as a consultant in most human cases of B virus in the United States since 1989.